CLASS TEST EDITION

INTERMEDIATE ALGEBRA

Models, Functions, and Graphs

Judith Kysh
University of California at Davis

Tom Sallee
University of California at Davis

Elaine Kasimatis
California State University, Sacramento

Brian Hoey
Christian Brothers High School

 **ADDISON-WESLEY**

An imprint of Addison Wesley Longman, Inc.

Reading, Massachusetts • Menlo Park, California • New York • Harlow, England
Don Mills, Ontario • Sydney • Mexico City • Madrid • Amsterdam

Reproduced by Addison Wesley Longman from camera-ready copy supplied by the authors.

Copyright © 1997 Addison Wesley Longman.

ISBN   0-201-84661-6

2 3 4 5 6 7 8 9 10 VG 99989796

PREFACE

This book is quite different from the mathematics books you have used in the past. First, it is all problems. Some are explorations; some are investigations; some are based on realistic situations; some are purely mathematical exercises. Many of the big ideas of the course are developed in and through the problems you will be doing during class working in small groups with other students with the assistance of your instructor. Only a part of the class time will be lecture. Much of the time will be allocated to working on the problems through which you and your group, assisted and guided by your instructor, will develop and verify mathematical generalizations and rules.

The approach of this text is to develop the important ideas over time, rather than covering them exhaustively in one chapter, testing and moving on. The new material introduced in early chapters is developed and used throughout the course. So at the end of a particular chapter you may not think you completely understand the material presented in that chapter, but several chapters later, after you've had the opportunity to use those ideas in a continuing sequence of problems, you should realize a deeper understanding than you could possibly achieve in the short term. Problems designed for assignments to be done outside of class provide continuing review and practice of the work done in earlier chapters.

ACKNOWLEDGMENTS

First we want to thank the high school teachers who used these materials in their classrooms in the developmental stages, for their contributions to the high school version on which this text is based.

Second we want to recognize and thank the college instructors who agreed to class test the first rough adaptation of the high school version. Their contributions have not only helped to improve this current class test version but will be a major factor in the preparation of the preliminary version.

Lindsey Bramlett-Smith.................................Santa Barbara City College; Santa Barbara, California
Jon Compton ....University of New Mexico; Los Lunas, New Mexico
Sarah Donovan ....Solano Community College; Suisun, California
Maggie Flint.....................Northeast State Technical Community College; Blountville, Tennessee
Dorothy Hawkes ....Solano Community College; Suisun, California
Brenda Jinkins ....State Technical Institute; Memphis, Tennessee
Kenneth Johnson..Sierra College; Rocklin, California
Roberta Lacefield...Waycross College; Waycross, Georgia
Derek Lance ....Chattanooga State Technical Community College; Chattanooga, Tennessee
Mike Mallen ....Santa Barbara City College; Santa Barbara, California
Karla Martin...................................... Walters State Community College; Morristown, Tennessee
Elizabeth Mefford.............................. Walters State Community College; Morristown, Tennessee
Dick Phelan ....Sierra College; Rocklin, California
Jim Ryan... Madera Center Community College; Madera, California
Karen Wootton ....Indiana State University; Terre Haute, Indiana

Finally we want to thank, in advance, the college instructors who will be using this class test version, for their willingness to work with the text while it is still undergoing significant revision and for their anticipated suggestions for making it better.

INTERMEDIATE ALGEBRA
TABLE OF CONTENTS

INTERMEDIATE ALGEBRA
Outline of Mathematics Content

All chapters include problems that require students to practice, use, or extend what they have worked on in previous units; therefore, only the material that is <u>introduced</u> by a unit is listed below.

REMINDER: A ll chapters include problems that require students to practice, use, or extend what they have worked on in previous units; therefore, only the material that is <u>introduced</u> by a unit is listed below.

WHAT DO THE BARS REPRESENT?

The bars on the cover for each chapter represent the six main threads of the course, six major areas in which you should be making progress throughout the course. At the beginning of each new chapter take a look at the each of the bars and ask yourself, "How am I doing? What are some of the things I now know for sure in this area? What do I need to do some more work on?"

Problem Solving: is a major concern of this course. You should be developing your ability to integrate your use of basic problem solving strategies such as guess-and-check and making organized lists with the use of algebraic procedures and your knowledge of graphing.

Representation/Modeling: Learning to represent situations with diagrams, graphs, equations or other models has become a much more important aspect of mathematics since the development of technologies which can carry out algorithmic procedures once a situation is represented. Being able to represent situations in several ways and to move back and forth between representations such as graphs, tables, and equations is emphasized throughout this course.

Functions/Graphing: This course is mostly about functions: linear, exponential, quadratic, other polynomial, logarithmic, rational, radical, and absolute value plus others that arise out of specific situations. To learn about functions you'll also need to know about some non-functions such as circles and "sleeping" parabolas.

Intersections/Systems: Intersections (and non-intersections) of graphs of functions (and non-functions) lead to systems of equations (and inequalities) which sometimes can be solved by algebraic means and sometimes cannot. You will build on what you know about solving linear and quadratic equations to solve 3x3 systems of linear equations, 2x2 pairings of lines, circles, parabolas and other polynomials that can be solved algebraically, and some pairings of these with exponential, and logarithmic equations for which you will have to use estimation and graphing or some combination of methods.

Algorithms: As you gain familiarity through practice with some types of problems that appear routinely as subproblems in other problems, you will need to develop facility with some routine procedures, including the tried and true tools of algebraic reorganization, rules of exponents, and others that you will identify as you work through the course. Procedures that you should learn so well they are almost automatic are identified throughout the course, often with a note to include them along with examples in the your **Tool Kit.**

Reasoning/Communication: includes developing the ability to give clear explanations, to make conjectures, and to develop logical mathematical arguments. Throughout the course you will practice articulating and justifying ideas orally, in writing in both mathematical and standard language, in diagrams, and sometimes based on a model.

Chapter 1
Exploring Functions
(Shrinking Arrows and Targets)

INPUT

The Fantastic
Function Machine

OUTPUT

PROBLEM SOLVING

REPRESENTATION/MODELING

FUNCTIONS/GRAPHING

INTERSECTIONS/SYSTEMS

ALGORITHMS

REASONING/COMMUNICATION

CHAPTER 1
SHRINKING ARROWS AND TARGETS: EXPLORING FUNCTIONS

INTRODUCTION

We assume that students who are taking this course are doing so to prepare for a future college course or technical field which will require that they be able to **use** the mathematics they have learned. The skills you learn in this course will be useful to you in many college courses, besides mathematics, such as chemistry, economics, psychology, and zoology. What this means is that

YOUR GOAL FOR THIS COURSE SHOULD BE UNDERSTANDING.

Only you will know if you understand something. We all know how easy it is to memorize something and even do well on a test without having any real idea about what is going on. If you settle for just performing well, you are cheating yourself, and when you need to use some part of the mathematics you have studied you will not have a clue as to where to begin.

We have done our best to design a course to make it easy for you to try to learn <u>and</u> to understand what you have learned. But no matter what -- learning is **hard work.** As the commercial says, NO (BRAIN) PAIN, NO GAIN. So if you want to do well in this course and get a good start in college there are three things which you will have to do:

1.   Make <u>understanding</u> the mathematics your highest goal.
2.   Discuss the questions with your group.
3 .   Do the homework.

If you want to understand, you will need to be willing to spend time thinking and trying out alternative approaches. Often you won't be able to come up with the right answer on the first try, so you need to be willing to stick with it. At this level there are generally several ways to think about a topic, and you should try to see more than one of them. That is one reason why we want you to work in groups.

A second reason for working in groups is that most job situations today demand that you work in groups, discussing ideas, listening, testing, taking the good parts of one person's idea and putting them with the good parts of someone else's idea to get a solution. Many of the problems in this book will ask you to discuss your ideas with your group and to listen to other people's ideas. This is an important practice to learn so do not skip over that part of the assignment.

A third reason, which is probably the most important, is that we want the mathematics to make sense to **you.** The mathematical techniques should not seem random. That is why the problems in this book have been structured so that you and your group, with some support from your instructor, can construct, and therefore understand, much more mathematics than you could from being given a rule and assigned a bunch of exercises which all look the same.

Finally, you need to do your homework. No one expects basketball players to become good (or even decent), if they just watch others play. The person has to get in there and practice. But they also need to know what to practice. In mathematics, you will often get stuck and not know what to do when you have no friend there and have no one to call to ask for help. In that situation write down what you do know about the problem, write down what you tried, and figure out what your question is. Then write down your question about the problem. This should be enough to convince your instructor that you have really tried to do the problem.

Doing this kind of work is what we mean by doing homework. In fact, the homework questions to which you should give the most attention are the ones you are not sure how to do. The questions which you can easily answer do not help you learn anything new; they are just practice. Getting stuck on a problem is your big opportunity to learn something. Analyze the problem to find out just what the hard part is. And then when you do find out how to do the problem, ask yourself, "What was it about that problem that made it hard?" Answering that question will get you ready to handle the next difficult problem so maybe you won't get stuck, and you will have learned something.

A very useful way to help yourself learn is to use such problem solving strategies as find a subproblem, guess and check, look for a pattern, organize a table, work backwards, use manipulatives, draw a graph, write an equation, or find an easier related problem. These strategies are not only useful for solving problems, but for **learning** as well. Much of what you learn in this course will not be brand new, but will build on mathematics which you already know, even if it has a new name. So you can use one of the problem solving strategies which we review in this first chapter to try to go from what you know to what you need to know.

Three of your major goals for this course should be: to become a better problem solver, to become better at asking questions, and to become better at explaining.

In Chapter 1 you will have the opportunity to:

- review problem solving strategies and begin to see them as learning strategies.

- recall and use skills that you have learned in previous classes in the context of modeling and solving larger problems.

- establish good group work routines.

- start to learn what it means to investigate a function.

- become familiar with function notation and begin to see many familiar equations and graphs as functions.

By the end of this chapter you should have reviewed some basic algebra skills and be able to:

- solve equations.
- write the equation of a line using slope and **y**-intercept.
- find mathematical models to fit real life data.
- draw graphs of lines and parabolas.
- use the Pythagorean Theorem.
- set up ratios to solve problems.

EF-1.   **YARNS AND OTHER INTRODUCTIONS**

In your group, make each of the following shapes with the yarn provided by your teacher. Show the shapes to your teacher as you complete them:

a)   square   b)   5-pointed star   c)   tetrahedron

d)   square based   e)   octahedron   f)   cube
      pyramid

EF-2.   <u>Example of Using a Guess and Check Table to Solve a Problem</u>

The length of a rectangle is three centimeters more than twice the width. The perimeter is 60 centimeters. Use a Guess and Check table to find how long and how wide the rectangle is.

STEP 1   Start a table. Why is the width a good thing to guess?

Guess Width	

STEP 2   Make a Guess.

Guess Width	
10	

Test your guess by writing down all of the steps you take to check it. These steps are your column descriptions.

STEP 3   Calculate the length.

Guess Width	Length	
10	23	

>>> *PROBLEM CONTINUES ON THE NEXT PAGE*>>>

STEP 4   Find the perimeter. Remember, rectangles have two "width" sides and two "length" sides.

Guess Width	Length	Perimeter
10	23	$2 \cdot 10 + 2 \cdot 23 = 66$

STEP 5   Check the perimeter against 60 and identify it as correct, too high, or too low.

Guess Width	Length	Perimeter	Check 60 ?
10	23	$2 \cdot 10 + 2 \cdot 23 = 66$	too high

STEP 6   Go back to Step 2 and make a new Guess. Repeat this process until you find the solution.

STEP 7   Write a sentence stating the solution.

EF-3.   Solve the following problem by using a guess-and-check table similar to the one above. Make at least three guesses even if you guess it right sooner. Be sure to write a sentence stating the solution.

One number is five more than a second number. The product of the numbers is 3,300. Find the two numbers.

EF-4.   Now go back to the end of your table for problem EF-2 and guess "**x**." Fill in each column of the table and write an equation that represents this problem. Do the same for EF-3. Then see if you can solve the equations you wrote to get the same results.

EF-5.   Write an equation to help you solve the following problem (Doing a guess-and-check table might help here). A cable 84 meters long is cut into two pieces so that one piece is 18 meters longer that the other. Find the length of each piece of cable.

> To **graph an equation** or **draw a graph** means we expect you to use graph paper, scale your axes appropriately, label key points, and plot points accurately.

EF-6.   Graph each of the following equations on separate sets of axes. Note: if you do not remember any "short cuts" for graphing, you can always make a table, choose a number for **x**, figure out **y**, and plot the point. Three points should be enough for (a), (b), and (c), but you should plot at least five for (d).

a) $y = -2x + 7$

b) $y = \frac{3}{5}x + 1$

c) $3x + 2y = 6$

d) $y = x^2$

EF-7.   How is the graph in part (d) above different from the other three graphs? Explain.

a)   What in the equation of part (d) makes its graph different?

b)   Do you remember the name of the graph in part (d)? If not, when you get to class check to see if someone in your group does.

EF-8.   Nafeesa graphed a line with slope 5 and y-intercept of -2. What is the equation of her line?

EF-9.   Write down everything you remember about the equation $y = mx + b$. You should include what kind of graph the equation represents and what **m** and **b** represent. Be as thorough as possible. If you can't remember very much right now, write down what you do know and check with your group tomorrow. Tomorrow this will become the first entry in your **TOOL KIT**.

EF-10.   Find the error in this problem. Explain what the error is and show how to do the problem correctly.

$$3(x - 2) - 2(x + 7) = 2x + 17$$
$$3x - 6 - 2x + 14 = 2x + 17$$
$$x + 8 = 2x + 17$$
$$-9 = x$$

EF-11.   Solve each of the following equations. This should be review so we have included the answers for you to check.

a) $\frac{3}{x} + 6 = -45$ b) $\frac{x - 2}{5} = \frac{10 - x}{8}$ c) $(x + 1)(x - 3) = 0$

$[\ x = -1/17 \approx -0.059 \ ]$ $[\ x = 66/13 \approx 5.08 \ ]$ $[\ x = -1, \ 3 \ ]$

EF-12.   Uyregor has a collection of normal, fair dice. He takes one out to roll it.

a)   What are all the possible outcomes that can come up?

b)   What is the probability that a 4 comes up?

c)   What is the probability that the number that comes up is less than 5?

EF-13.   Read or re-read the introduction to the course before problem EF-1, focusing on the goals of this course. Write a paragraph that outlines your goals. Include information about how this course will assist you in fulfilling your goals. Be prepared to discuss this with your group.

EF-14. **KEEPING A NOTEBOOK**

> You will need to keep an organized notebook for this course. Here is one method. Check with your instructor as to whether you should follow these guidelines or whether there is some other system you should follow.
>
> • The notebook should be a sturdy three-ring lose-leaf binder with a hard cover.
>
> • The binder should have dividers to separate it into six sections:
>   TEXT
>   HOMEWORK/CLASSWORK
>   NOTES
>   TESTS AND QUIZZES
>   TOOLKITS
>   GRAPH PAPER
>   JOURNAL
>
> Your notebook will be your biggest asset for this course and will be the chief way to prepare for tests, so take good care of it!

EF-15. **INTRODUCTION TO INVESTIGATING A FUNCTION**

> As you begin investigating functions, it is important that you understand what we expect when we ask you to sketch a graph. To **sketch a graph** means you show the approximate shape of the graph in the correct location with respect to your axes and that you clearly label all key points.

Consider the following equation: $y = \sqrt{(4 - x)} - 1$ Use a graphing calculator to help you to answer each of the following completely.

a) Sketch the graph from your graphing calculator.

b) What are all the possible values of **x**? Are there any values that will not work for **x**? What is the largest value you can use for **x**? Explain.

c) Does the graph ever cross the horizontal line $y = 15$? How about $y = 500$? How do you know?

d) What is the smallest possible value of y? What are all of the possible values of y?

e) Have you found <u>everything</u> about this equation that is important? Hmmm...

f) Does the line $y = x$ intersect the graph? How could you find the point of intersection?

♀EF-16.   Refer back to problem EF-9 regarding a linear equation. On the Tool Kit sheet (copy at the end of this chapter), you will see several topics. Some of these topics should be familiar to you. Fill in the information that will be useful to you in understanding and remembering the topic of linear equations. We will fill in the other topics later in this chapter and during Chapter 2.

EF-17.   Write down everything you can remember about the term **intercepts**. Discuss what you remember with your group and decide what is important. Be sure you include examples.

EF-18.   Solve each of the following for **x**.

   a)   Forty-two percent of **x** is 112.

   b)   Forty-two is **x** percent of 112.

   c)   Twenty-seven is **x** percent of 100.

   d)   Twenty-seven percent of 500 is **x**.

EF-19.   In each of the following equations, what is **y** when x = 2? When x = 0? Where would the graph of each cross the **y**-axis?

   a) $y = 3x + 15$

   b) $y = 3 - 3x$

EF-20.   Suppose we want to find where the lines $y = 3x + 15$ and $y = 3 - 3x$ cross and we want to be more accurate than the graphing calculator or graph paper will let us be. We can use algebra to find this point of intersection.

   a)   If you remember how to do this, find the point of intersection using algebra and explain your method to your group tomorrow in class. If you don't remember, then do parts (b) - (e) below.

   b)   Since $y = 3x + 15$ and $y = 3 - 3x$, what must be true about $3x + 15$ and $3 - 3x$?

   c)   Write an equation which contains no y's and solve it for x.

   d)   Use the x-value you found in part (c) to find the corresponding y-value.

   e)   Write the coordinates of the point where the two lines cross.

EF-21.   Find the error in the problem. Explain what the error is and re-do the problem showing
          a correct solution.

$$\frac{5}{x} = x - 4$$

$$x \bullet \frac{5}{x} = x - 4$$

$$5 = x - 4$$

$$9 = x$$

EF-22.   The perimeter of a triangle is 76 centimeters. The second side is twice as long as the
          first side. The third side is four centimeters shorter than the second side. Write an
          equation and find the length of each side.

EF-23.   Solve each of the following equations:

   a) $3x + 5 = 8x - 72$

   b) $7x + 2(3x + 1) = 5(6x - 3)$

   c) $6x - 7(x + 4) = 5 + 2(x + 3)$

EF-24.   In $\triangle ABC$ shown below, $\overline{PQ}$ is drawn x units down on side $\overline{AB}$, and is parallel to $\overline{BC}$.

          $\overline{AB}$ has a length of 10, $\overline{BC}$ has a length of 8.

   a)   If x = 1, what is the length of y?

   b)   Suppose x = 2. Now what is the length of y?

   c)   Suppose x = 3. Now what is the length of y?

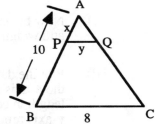

   d)   Write an equation relating x and y. Write it in the form "y = ."

   e)   Use your equation to find y when x = 7. Explain in words what this means.

   f)   If you graphed your equation, what do you think the graph would look like?
         Explain how you know.

   g)   The relationship described in this problem is often called direct variation. Why
         does this description make sense?

SHRINKING ARROWS LAB
(EF-25 to EF-31)

In this investigation you will be comparing weights of pencils sharpened to different lengths. While this activity, on the face of it, may not seem very inspiring or potentially enlightening, we have chosen it because it will lead to the development of mathematical models and concepts that are central to what you will be learning in this course. (A "mathematical model" might be new to you. One possible model is an algebraic equation(s) used to represent a real world situation. Knowing an equation that represents a situation allows us to make predictions based on the model.) Although this activity does not quite represent a real-world situation, it is an investigation that a large group of people can do fairly quickly in a classroom, a fact that will allow you to focus on the mathematical model this experiment will generate.

Be sure that each person in your group does a neat and accurate graph, answers all questions, and shows all work. Each of you will be keeping a portfolio in this course which will show examples of your work and problem solving skills. Your lab write-ups to these investigations will be the first pieces you will put in your portfolio.

EF-25.   In this investigation, you are going to sharpen pencils! Well, OK, you will do some math too, but throughout this investigation, remember not to use the eraser for erasing! Here's how it works:

a)   Start with a fresh pencil. Sharpen it just enough to be usable. Measure the length of the painted part (excluding the metal and eraser part) in centimeters and then weigh the pencil. Record both quantities in a table with two headings "Length of painted part (cm)" and "Mass (g)."

b)   Before you begin gathering further data, do you think the relationship will be linear or non-linear? Why?

c)   Now break off the point and sharpen the pencil a bit. Measure the painted part and weigh it. Record your data in your table.

d)   Plot the data on a graph where the x-axis represents the length in centimeters and the y-axis represents the mass in grams. Use a full sheet of graph paper.

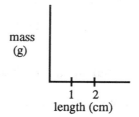

e)   What should happen to the mass of the pencil as you sharpen it. How does this relate to the graph? Be clear. Collect at least ten data points, and the last few should be with the pencil down to a little nub. Don't forget to measure, weigh, and record after each sharpening!

f)   Graph your data on your graph paper. Write a sentence comparing the graph to your predictions.

EF-26.   Draw a line to fit your data and find the equation of this line that "best fits" the graph of your data.

EF-27. Suppose you sharpened the pencil all the way down, so that no paint remained.

   a) How much would this little pencil weigh?

   b) Do you need to sharpen the pencil all the way down and weigh it to answer this question? Or can you predict from your graph?

   c) Explain your answer to part (b).

EF-28. Use your line-of-best-fit to approximate the change in mass per 1 cm piece of pencil. What would this value represent in your linear equation?

EF-29. Answer each of the following in complete sentences.

   a) Explain what the **y**-intercept and slope of the line correspond to on the pencil.

   b) Explain why the data points for this lab lie in a straight line.

   c) Think of and describe a different situation (that doesn't involve weighing) that would produce data points that would lie in a straight line.

   d) Explain what your new situation has in common with weighing a pencil.

EF-30. The **x**-axis on your graph represents possible lengths for the painted part of the pencil. What are acceptable values for **x** in this problem? Explain.

EF-31. What do the **y**-values represent? What are possible values?

   Be sure to save your data and the answers to the previous questions for this investigation for your lab report. The purpose of this investigation and the next one you will do is to show how you can create an algebraic representation for the patterns you find in the data you collect.

EF-32.   Teller K. Draper completed the tables and drew the graphs below but his teacher
          marked each one wrong. Be as specific as you can when you answer the questions
          below. For each graph:

a)   Identify as many mistakes as you can.

b)   Describe in a sentence what to do to correct each mistake.

c)   Redraw the graphs so that Teller knows what to do and can get full credit next
      time.

i) $y = 2x + 1$

x	y
1	3
2	5
3	7

ii) $y = x^2$

x	y
0	0
1	1
2	4
3	9

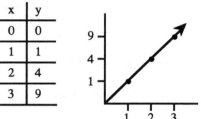

iii) $y = 3x + 50$

x	y
0	50
1	53
2	56

iv) $y = x^2 - 4$

x	y
0	-4
-1	-3
1	-3
2	0
3	5

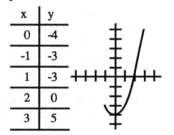

EF-33.   Rearrange these equations by solving each of them for **x**. Write each equation as
          x = _____ (y will be in your answer).

a) $y = \dfrac{3}{5}x + 1$ b) $3x + 2y = 6$ c) $y = x^2$

EF-34.   What value of **x** allows you to find the **y**-intercept? Where does the graph of each of
          these cross the **y**-axis?

a) $y = 3x + 6$ b) $x = 5y - 10$

c) $y = x^2$ d) $y = 2x^2 - 4$

e) $y = (x - 5)^2$

EF-35.   What value of **y** allows you to find the x-intercept? For each of the equations in parts
          (a) - (e) above, find where its graph intersects the x-axis.

EF-36   Imagine that we add water to the beakers, A, B, and C, shown at right. Sketch a graph to show the relationship between the volume of water added and the height of the water in each beaker. Put all three graphs on one set of axes (you may want to use colored pencils to distinguish the graphs). What are the independent and dependent variables?

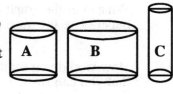

EF-37.   Write each of the following as a single fraction:

a) $\dfrac{3}{8} + \dfrac{5}{12}$ b) $\dfrac{1}{6} + \dfrac{2}{a}$

c) $\dfrac{1}{b} + \dfrac{2}{a}$ d) $\dfrac{3}{c} - \dfrac{5}{d}$

EF-38.   Tony says that $(x + y)^2$ is the same as $x^2 + y^2$ but Patrice says Tony is wrong. Who do you agree with? Explain in at least two different ways (using numbers, algebra, diagrams or other approaches) why the person you agree with is right. You have to convince the other, who is very strong minded.

EF-39.   Stacie says to Cory, "Reach into this standard deck of playing cards, and pull out any card at random. If it is the queen of hearts, I'll pay you $5.00." What is the probability that Cory gets Stacie's $5.00? What is the probability that Stacie keeps her $5.00? Justify your answers.

USING A GRAPHING CALCULATOR:
(EF-40 to EF-44)

EF-40.   Take out your solutions to EF-33, EF-34, and EF-35 from your last assignment. Be sure to clear all the old stuff off your calculator, then check that your intercepts are correct by using the graphing calculator to display the graphs. Considering the following questions may help in answering EF-33(b), EF-34(d), and EF-(e).

a)   What did you have to do with the equation in EF-33(b) in order to graph it?

b)   For EF-35(d) use the trace button to check the x-intercepts. How close did you get to y = 0? Try the zoom button. How close did you get to the actual x-intercepts?

c)   What window adjustment did you have to make in order to see the y-intercept in EF-34(e)?

EF-41.   Where do the graphs of y = 2x - 6 and y = -4x + 6 cross? Use your graphing calculator to find the point. Use your zoom key to get closer.

EF-42.   Sketch the graphs of y = 3x - 5 and y = 2x + 12. Adjust the window and use your zoom in feature to find our where they cross. Verify your answer by solving algebraically.

EF-43.   What does the graph of $y = x^2 - 3x - 3$ look like? What do we call it? Make a sketch on paper. Use the trace and/or zoom buttons to approximate the x-intercepts and vertex.

♀EF-44.   How would you set the window on the graphing calculator to graph $y = (x + 1)(x - 9)$ so that you can see the **complete** graph?   By **complete**, we mean that we see everything that is important about the graph and that everything off the screen is predictable based on what we do see. Another way to think about it is to say a graph is complete when further zooming or window changes will not add any new information. Explain and check.

♀EF-45.   Does the temperature outside depend on the time of day, or does the time of day depend on the temperature outside? This may seem like a silly question, but to sketch a graph that represents this relationship, we need to determine which axis will represent what quantity.

   a)   When you graph an equation such as $y = 3x - 5$, as you did earlier, which variable, the **x** or the **y**, **depends** on the other? Which is not dependent, that is, which is **in**dependent? Explain.

   b)   Which is <u>dependent</u>: temperature or time of day? Which is <u>independent</u>?

   c)   Sketch a graph, with appropriately named axes, showing the relationship between temperature outside and time of day.

This notion of independent and dependent variables is an important concept. Later we will ask you to write the important ideas of this chapter down. Don't forget about these terms when we ask you to do it!

EF-46.   Examine each graph below. Based on the shape of the graph and the labels of the axes, describe the relationship that each graph represents. State which is the independent and which is the dependent variable.

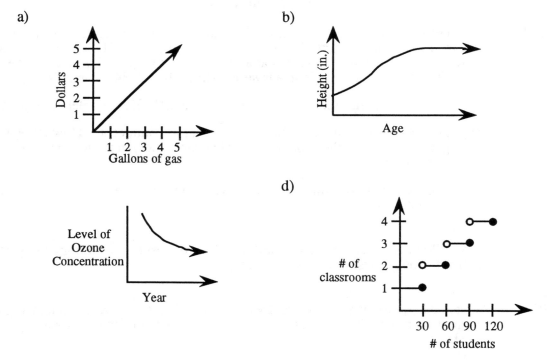

a)

b)

d)

EF-47.   Find where the following pairs of lines intersect.

a) $y = 5x - 2$
      $y = 3x + 18$

b) $y = x - 4$
      $2x + 3y = -17$

EF-48.   Find the error in the problem below. Identify the error and show how to do the problem correctly.

$$4.1x = 9.5x + 23.7$$
$$\underline{-4.1x \quad -4.1x}$$
$$5.4x = 23.7$$
$$\frac{5.4x}{5.4} = \frac{23.7}{5.4}$$
$$x = 4.39$$

EF-49.   Factor and solve these quadratic equations:

a) $x^2 - 7x + 10 = 0$

b) $x^2 + x - 42 = 0$

c) $3x^2 + 15x = 0$

d) $2x^2 - x = 3$

EF-50.   Tracy is convinced that $x^3 + x^2$ must equal x^5. Your job is to convince her she is mistaken. Use at least two different ways to demonstrate to Tracy why $x^3 + x^2 \ne x^5$.

EF-51.   Ivan overheard part of your discussion with Tracey and said, "I know why that can't be right. It's because $(x^3)^2 = x^5$." Now explain in at least two different ways to Ivan why he is also mistaken.

EF-52.   There is a combination of powers of 2 and 3 with x as a base that does give the answer x^5. What is it? Show why this is true.

EF-53.   An average school bus holds 45 people. Sketch a graph showing the relationship between the number of school buses and the number of people they can transport. Be sure to label the axes.

DOMAINS AND RANGES

EF-54.   Look back to your first function investigation, EF-15. Describe all the x-values that could be used in the equation. Were there any restrictions? Explain.

EF-55.   Find your data from the shrinking arrows lab. What were the possible values for the lengths of the arrows? Could there be any other values that you did not include? Explain.

EF-56.   For the arrow lab what is the independent variable: the length or the weight?

♀EF-57.   Include the following definition in your Tool Kit:

> The set of possible values that the **independent variable** can take on (input values) has a special name. It is called the **domain** of the function. It consists of every number **x** can represent for your function.

EF-58.   Again look back to your first function investigation, $y = \sqrt{(4 - x)} - 1$. What were the possible **y**-values you found for the equation? What did **y** depend on? Explain.

EF-59.   What were the values for the weights of the arrows? Did the weight of the arrows depend on anything? Explain.

EF-60.   For the arrow investigation, what is the dependent variable: the length or the weight?

♀EF-61.   Add this definition to your Tool Kit:

> The **range** of a function is the set of possible values of the **dependent variable** (output values). It consists of every number **y** can represent for your function.

EF-62.   Use a set of axes to represent the domain and another to represent the range for each of the following functions.

a)   b)   c)

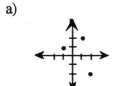   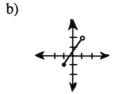   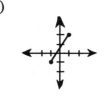

EF-63.   Graph these two lines on the same set of axes: $y = 2x$ and $y = -\frac{1}{2}x + 6$.

a)   Find the **x** and **y** intercepts for each equation.

b)   Shade in the region bounded by the two lines and the x-axis.

c)   What is the range and domain of the region?

d)   Find the area of this region, accurate to the nearest tenth.

e)   What is the angle between the two lines? There is a word that describes the relationship between the two lines. What is it?

f)   What do you notice about their slopes?

EF-64.   In the last problem, did you need to find the point where the two lines intersected?

   a)   Explain how to find the point of intersection for the two lines in the previous problem.

   b)   Is using the graph an accurate way to find the point of intersection? Explain.

   c)   What method do you think is the most accurate? Explain.

EF-65.   Nissos and Chelita were arguing over a math problem. Nissos was trying to explain to Chelita that she had made a mistake in finding the x-intercepts of the function $y = x^2 - 10x + 21$. "No way!" Chelita said. "I know how to find x-intercepts! You make the y equal to zero and solve for x. I know I did this right!" Here is her work.

$$x^2 - 10x + 21 = 0$$
$$(x + 7)(x + 3) = 0$$

Therefore:

$$x + 7 = 0 \text{ or } x + 3 = 0$$
$$x = -7 \qquad x = -3$$

Nissos tried to explain to her that she still had done something wrong. Who is correct? Justify and explain your answer completely.

EF-66.   In Salem, California, 42% of the registered voters are Republicans and 44% are Democrats. The other 217 are Independents. How many registered voters are there in Salem? Write an equation that represents this problem.
Suggestion: You might want to set-up a guess and check table first to find the solution. Then try writing the equation.

EF-67.   For each of the following, state the domain and range.

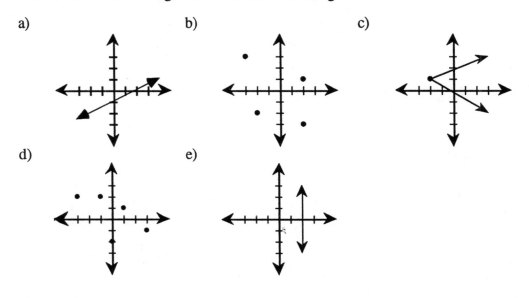

a)   b)   c)

d)   e)

EF-68. Solve each of the following equations for **x**.

   a) 1234x + 23456 = 987654

   b) $\dfrac{10}{x} + \dfrac{20}{x} = 5$

   c) $x^3 - 3x^2 + 2x = 0$ (Remember you can factor out a common factor.)

   d) $5x^2 - 6x + 1 = 0$

EF-69. A circle has an area of 45 cm^2. Find its circumference (perimeter). Remember $C = 2\pi r$ and $A = \pi r^2$.

   a) List all the STEPS you will need to complete to solve this problem.

   b) These STEPS are known as **SUBPROBLEMS**. Keep this term in mind.

   c) Solve the original problem. Give your answer exactly (here it will involve π and a square root) then use your calculator to approximate it to two decimal places.

SUBPROBLEMS

Most problems which you will meet in this course, in other mathematics courses, and in life, are made up of smaller problems whose solutions you need to put together in order to solve the original problem. These smaller problems are called SUBPROBLEMS. For example, you cannot solve the next problem directly. First you must solve the subproblems which are not stated explicitly but which are necessary to solve in order to solve the larger problem.

EF-70.   The Logan's dog, Digger, was constantly digging his way under the fence, escaping from the back yard, and making a nuisance of himself in the neighborhood. Digger was also in big trouble with Connie Logan who enjoyed her leisure time away from her law firm by taking care of her flower garden. They decided they would have to keep Digger tied up while they were away during the day. After a long discussion about how much of the back yard Digger should be allowed to destroy, they agreed that at least **two-thirds** of the yard should be safe for planting flowers and other plants. They tied Digger to the corner of the shed, at point A, with a **25 foot** rope. Will this satisfy Connie?

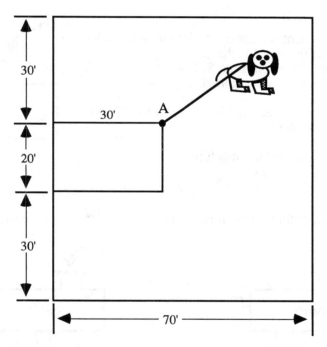

a)   Draw the region in which Digger is free to roam.

b)   Before actually working out any answers, write down all of the subproblems you need to solve the problem.

c)   Now solve each subproblem and answer the question.

EF-71.   How will your solution to the previous problem change if we lengthen Digger's rope? Explain.

EF-72.   Carmichael B. Pierce, mathematician extraordinaire, has invented a fascinating machine that he hopes to sell to the Logans to cut down on the amount of work they are doing in calculating the destruction area of their dog Digger. All you have to do is drop in a specific rope length, and the machine works out the area.

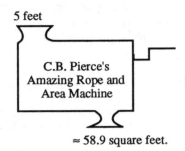

5 feet

C.B. Pierce's
Amazing Rope and
Area Machine

≈ 58.9 square feet.

If this machine **really** works and Connie puts in a specific rope length, is it possible for two different answers to fall out? That is, if she puts in 25 feet today, and puts in 25 feet again two weeks from today, will it give her the same answer? Explain.

EF-73.   Terri's assignment for her math class was to make up rule for a Function Machine. If we put a 3 into her machine, the output is an 8. If we put in a 10, it gives us 29, and if we put in 20 it gives us 59.

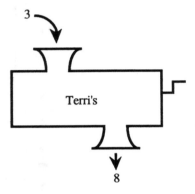

a)   What would her machine do to 5? to -1? to x? A table may help.

b)   Write a rule for her machine.

EF-74.   Gerri made a different machine. Here are four pictures of the same machine. Find its rule.

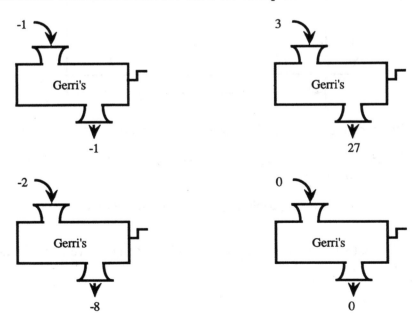

EF-75.   Carmichael B. Pierce also made a different function machine. The inner "workings" of the machine are visible. What will be the output if:

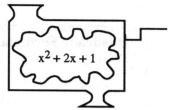

a)   3 is dropped in?

b)   -4 is dropped in?

c)   -22.872 is dropped in?

EF-76.   If the number 1 is the output for Carmichael's machine, how can you find out what number was dropped in? Find the number or numbers that could have been dropped in.

EF-77.   Find the error in the problem below. Describe the error, and show how to solve the problem correctly.

$$x^2 - 10x + 21 = 5$$
$$(x - 7)(x - 3) = 5$$
$$x - 7 = 5 \text{ or } x - 3 = 5$$
$$x = 12 \qquad x = 8$$

EF-78.   Write an equation or two equations to help you solve the following problem. A rectangle's length is four times its width. The sum of two consecutive sides is 22. How long is each side?

EF-79.   Consider the equation $4x - 6y = 12$.

a)   What do you suppose the graph of this equation looks like? Justify your answer.

b)   Solve the equation for y and graph the equation.

c)   Explain how to find the x- and y-intercepts. Be complete so that you can give your explanation to someone who is still having trouble with intercepts and he or she will be able to follow and understand your explanation.

d)   Which form of the equation is best for finding intercepts quickly? Why?

e)   Use the intercepts you found in part (c) to graph the line. Did you get the same line you got in part (b)? Should you? Explain.

♀EF-80.   **THE QUADRATIC FORMULA**

> An important tool from Algebra 1 that you might have already used this year is the **Quadratic Formula**. This formula is particularly helpful in finding the intercepts of parabolas. As you probably remember, the formula states
>
> $$\text{If } ax^2 + bx + c = 0, \text{ then } x = \frac{-b \pm \sqrt{b^2 - 4ac}}{2a}.$$
>
> For example, suppose we wanted to find the x-intercepts of $y = 2x^2 - 3x - 3$. First we would let $y = 0$ (write an explanation in your Tool Kit for why we let $y = 0$),
>
> $$0 = 2x^2 - 3x - 3$$
>
> then, since this does not factor, we must use the Quadratic Formula to solve for **x**.
>
> First identify **a, b**, and **c**. $a = 2$, $b = -3$, $c = -3$.
>
> $$x = \frac{-(-3) \pm \sqrt{(-3)^2 - 4(2)(-3)}}{2(2)}$$
>
> $$= \frac{3 \pm \sqrt{9 + 24}}{4}$$
>
> $$= \frac{3 \pm \sqrt{33}}{4}$$
>
> $$\approx \frac{3 \pm 5.745}{4}$$
>
> $$\text{So, } x \approx \frac{3 + 5.745}{4} \text{ and } x \approx \frac{3 - 5.745}{4}$$
>
> $$x \approx 2.18625 \text{ and } x \approx -0.68625.$$
>
> Be sure you include all the information you need about the **Quadratic Formula** in your Tool Kit

EF-81.   Now solve these equations and give the solution in both radical and decimal form.

a) $x^2 - 5x + 3 = 0$ b) $x^2 + 3x - 3 = 0$

c) $3x^2 - 7x = 12$ d) $2x^2 + x = 6$

EF-82.   Have you ever wondered why so many equations are written with the variables **x** and **y**? Even if you haven't, suppose you were going to reach into a bag that contained the alphabet, and pull out one letter at random to use as a variable in equations. What is the probability that you would pull out an **x**? If you got the **x**, now what is the probability that you pull out the **y**?

TARGET LAB
(EF-83 to EF-85)

EF-83.   Get a sheet of cardboard. Find and mark the <u>exact</u> center. With a compass, make and cut out the largest circle you can from the cardboard. Then draw about nine more concentric circles on the cardboard (<u>not</u> equally spaced), but <u>don't</u> cut them out yet! Set up a table with headings, "Length of radius (cm)" and "Mass of disc (g)." Record the length of the radius in the first column. You are going to cut out each circle starting from the outside, and weigh the disc of cardboard each time, recording the information in a table. When your table is complete, you will graph the results.

   a)   Before you graph the results, what do you think the graph will look like?

   b)   Graph your results and sketch in a "best fitting" curve.

   c)   What kind of curve does your graph seem to be?

   d)   Predict the mass of a circle with a radius twice as large as your largest circle. Explain how you figured this out.

EF-84.   For <u>your</u> piece of cardboard, what is the domain of **x**-values? (That is, what are acceptable lengths for the radius?) What is the range for **y**-values?

EF-85.   Does your graph have **x** or **y** intercepts? If so, what are they and what do they represent? If not, explain completely why not.

   REMEMBER! You will need your data from this investigation for your lab reports. Your instructor will tell you what format to use.

EF-86.   Remember Logan's dog Digger? Suppose Digger is in the middle of an enormous yard, the yard contains acres and acres of land. Digger is roped to a stake at the center of the yard. Consider a function where the input is the length of Digger's rope and the output is the amount of area that Digger has to roam in. Write an equation for this relationship and sketch its graph.

EF-87.   Find the domain and the range for each of the following functions.

   a)   b)   c)   d)

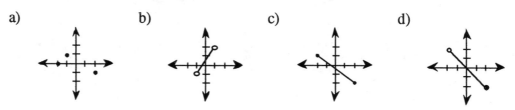

EF-88.   Recently, Kalani and Lynette took a trip
          from Vacaville to Los Angeles. The graph
          at right represents their trip.

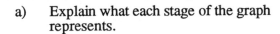
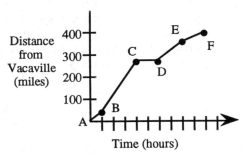

          a)   Explain what each stage of the graph
                represents.

          b)   About how many miles is it from
                Vacaville to Los Angeles? How do
                you know?

          c)   Challenge: Using the graph above, sketch a graph that would represent their
                speed while traveling. Be sure to label axes.

EF-89.   If $t(n) = n + 3$, what is:

          a) $t(1)$?   b) $t(100)$?

EF-90.   Write an equation for the line passing through the points (2, 0) and (0, -3). Remember
          that drawing a diagram, in this case the graph, can be very helpful.

EF-91.   Make a sketch of a graph showing the relationship between the number of hours a
          student studies for a test and the student's score on the test.

EF-92.   Show two ways to solve for **x**: $7x^2 - 3x - 4 = 0$.

EF-93.   I noticed my outdoor faucet dripping even after I tried to turn it off. So I decided to see
          how much water was being wasted. I put a glass under it at 7:33 p.m. and removed it
          at 9:19 p.m. When I poured the water out of the glass into a measuring cup, it
          measured almost exactly $1\frac{1}{3}$ cups of water. About how much water leaks from that
          faucet each day? What subproblems did you do to solve this?

EF-94.   Write an equation that represents the relationship between the number of days the faucet
          has been dripping and how much water has been wasted.

          a)   What are the independent and dependent variables?

          b)   What are the domain and range?

          c)   What should the graph look like? Justify your answer.

EF-95.   Use what you know about similar triangles to complete each
of the following ratios.

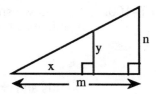

$$\frac{y}{x} = \underline{\hspace{1cm}}$$

$$\frac{n}{y} = \underline{\hspace{1cm}}$$

EF-96.   **INVESTIGATING A FUNCTION**

Working together, find out everything you can about the following equation:
$y = \sqrt{(x + 16)}$. Be ready to report your group's findings in about 15 minutes. (By
the way, part of the problem is for your group to decide what we mean by "everything
there is to know." How do you know you have found everything? Keep asking,
"What else could we check?") Be sure to save your work as you will need it again.

EF-97.   When you use a graphing calculator for this problem, you need to be careful when
typing in the equation of the function. The calculator follows the same order of
operations you have learned in earlier math courses, so for the graph below, do you
want to take the reciprocal of just the **x** or of the **x - 3**? What can you do to make the
calculator do what you want? Be sure to sketch your graph on your paper.

$$y = \frac{1}{x - 3}$$

EF-98.   You can use the graphing calculator to make the kind of calculations you did in EF-75
part (c) easier. As we saw, that function machine squared the number that was dropped
in, added it to twice the number dropped in, and finally added one to all that. If -3 is
dropped in, what comes out? Now take advantage of the graphing calculator and
calculate the result when 23 is dropped in.

EF-99.   A convenient way to show what a function machine does is to use **function
notation**. For Carmichael's machine in EF-75, we would write $f(x) = x^2 + 2x + 1$.
The **f** is just the name of the function machine, it is not a variable. It could just as well
be Pierce$(x) = x^2 + 2x + 1$ if the machine happened to be named Pierce! In part (c) of
EF-75, you actually found f(-22.872). Use function notation to describe what each of
the following machines does to **x**.

a)

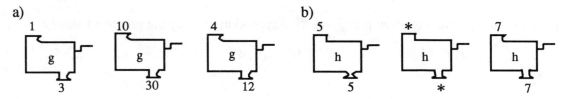

c)

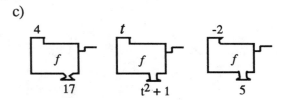

EF-100.   Now, how can we use the graphing calculator to make substituting values easier? Just
           type in exactly what you want to find. For instance, to find f(-22.872) type in

$$(-22.872)^2 + 2(-22.872) + 1$$

           Do this and check your answer.

EF-101.   Using the same function, find f(-142.2). You can make your work even easier by
           substituting values on your graphing calculator and just changing the input value from
           -22.872 to -142.2. Be prepared! Different calculators work different ways. Try this
           and check your result with your group.

EF-102.   Compare each of the following tables.

           $f(x) = x^2 + 2x + 1$ $y = x^2 + 2x + 1$

x	f(x)
-22.872	478.384384
-142.2	19937.44

x	y
-22.872	478.384384
-142.2	19937.44

           For each table, what is the independent and dependent variable? What is the difference
           between the two tables? What is the difference between the two rules? Explain.

EF-103.   If $g(x) = .01x^2 - 6.03x + 17.1$, find:

           a)   g(-5.1)   b)   g(6.23)

EF-104.   Graph the following and find the **x**- and **y**-intercepts.

           a) $y = 2x + 3$ b) $f(x) = 2x + 3$

EF-105.   Is it true that $\frac{x + 2}{x + 3} = \frac{2}{3}$? Explain your reasoning.

EF-106.   Which ordered pair gives the expression 5x + 2y the greatest value?

           a)   (4, 6)   b)   (6, 4)   c)   (-6, -5)

EF-107. For each of the functions below, what are the domain and range?

a)

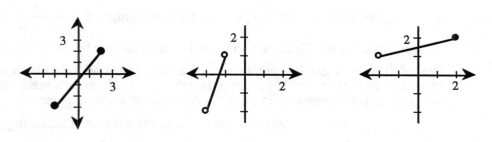

EF-108. From each of the following pairs of domain and range graphs, sketch two different graphs of functions.

a)   b)

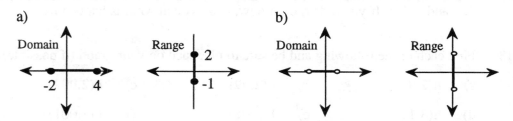

EF-109. Solve each of the following equations for the indicated variable.

a) $y = mx + b$ (for **x**)   b) $A = \pi r^2$ (for r)

c) $V = LWH$ (for W)   d) $2x + \dfrac{1}{y} = 3$ (for y)

EF-110. At right are the side views of three swimming pools. Below are three graphs that show the relationship between the depth of the water (in the deep end) over time as the pool is filled at a constant rate. Match each of the following swimming pool profiles, A, B, and C, to the graph that could represent this relationship.

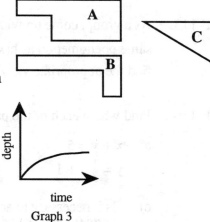

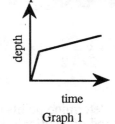

depth   depth   depth

time   time   time

Graph 1   Graph 2   Graph 3

INVESTIGATING FUNCTIONS WITH ASYMPTOTES

EF-111.   In the last assignment, in EF-97, you sketched the graph of $y = \dfrac{1}{x - 3}$ using your

graphing calculator.   Today we will consider the function $h(x) = \dfrac{1}{x - 3}$ in more detail.
Use a full sheet of graph paper, and set up a pair of axes with the origin near the center,
vertically, and about 2 inches in from the left side. On both axes make each square on
the graph paper represent 0.25 units. Now make an accurate graph of

$h(x) = \dfrac{1}{x - 3}$. Use a calculator to make the substituting and calculating easier, but

make an extensive table, do not just copy the picture of the graph.

EF-112.   What seems to be happening to the graph at x = 3? Did you use values like 3.5, 3.25,
2.5, and 2.75? If you didn't, do it now, and explain what is happening.

EF-113.   Find each of the following and be sure to plot each on your graph (if possible).

a)   h(2.9)   b)   h(2.99)   c)   h(2.999)

d)   h(3.1)   e)   h(3.01)   f)   h(3.001)

g)   What happens, on the graph and algebraically, when you try h(3)? Give a good
explanation for this phenomenon.

EF-114.   a)   What are the domain and range for the function h? Explain.

b)   Where does the graph of **h** cross the x-axis? Compare this to your picture and
justify your answer.

EF-115.   As a group, come up with another equation that will have a different graph, but the
same phenomenon as $h(x) = \dfrac{1}{x - 3}$. Test it out. Give it to another group to graph and
find the appropriate viewing screen to be able to see the complete graph.

EF-116.   Find where each of the pairs of graphs intersect.

a) $x + y = 5$ b) $2x + y = 3$
$\quad y = \dfrac{1}{3}x + 1$ $\quad y = -2x - 2$

c)   Is it necessary to actually draw the graphs to determine points of intersection in
parts (a) and (b) or is there an algebraic method of solution? If you didn't
already, show how you could solve each system of equations algebraically.

EF-117.   If $t(n) = 5 - 2(\frac{3}{4}n - 1)$, find:

a) $t(4)$ b) $t(-20)$ c) $t(\frac{2}{3})$

EF-118.   You make \$4.50 per hour at your job. Approximately 30% of your wages goes for taxes, union dues, and deductions. How many hours do you need to work to take home \$135 per week? Use equations or explain your reasoning.

EF-119.   If $h(x) = x^2 - 5$, where does the graph of $h(x)$ cross the x-axis? Make a sketch of the graph.

EF-120.   Solve the following for z.

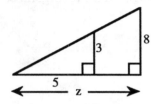

EF-121.   If $f(x) = -x^2 + 2x^3$, find:

a) $f(-1)$ b) $f(-2)$ c) $f(2)$

d) $f(1)$ e) $f(0)$

f)   Draw a graph of this function.

g)   What value(s) of **x** will make $f(x) = 0$?

EF-122.   Burt Balding takes a shower in a 40 gallon tub. The drain in his tub can drain up to 10 gallons of water per minute when the drain is clear. Unfortunately, Burt's rapid hair loss slows the drainage considerably! For each minute that Burt showers, he loses a set amount of hair and the drain becomes more and more clogged. Suppose that the water is flowing into the tub at a rate of 5 gallons per minute, that the water in the tub starts to back up after 10 minutes, and that Burt likes to take **long** showers. Make a reasonable graph of this situation showing how the gallons of water in the tub and the showering time (in minutes) are related. Label the axes, important points, and any other critical items.

EF-123.   Make a sketch of a graph showing the relationship between the number of people on campus and the time of day.

EF-124.   The graph below shows the probabilities of a certain spinner coming up a red, blue or green.

      a)   What is the probability the spinner comes up blue? b)   What is the probability the spinner comes up green? c)   Which axis represents the <u>independent</u> axis? The <u>dependent</u> axis?

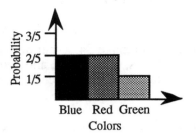

      d)   What is the domain? The range?

MORE INVESTIGATIONS

EF-125.   Look back at your work on investigating functions. Make an organized list of all the elements of investigating a function.

EF-126.   Investigate the function $y = \dfrac{5}{(x^2 + 1)} - 1$.

EF-127.   **CHAPTER 1 SUMMARY ASSIGNMENT**

There are a lot of ideas covered in this chapter. Some ideas should be review while others may be new to you. Answer the following questions completely.

      a)   Write down at least three major ideas of the chapter and find a minimum of one problem which illustrates each idea. Be sure to include a description of the original problem and a completely worked out solution.

      b)   If you had to choose a favorite problem from the chapter, what would it be? Why?

      c)   Find a problem that you still cannot solve or that you are worried that you might not be able to solve on a test. Write out the question and as much of the solution as you can until you get to the hard part. Then explain what it is that keeps you from solving the problem. Be clear and precise.

      d)   In this chapter you worked in your group daily. What role did you play in the group? (Discuss ideas similar to: were you a leader, taking charge? Did you keep your group on task? Did you ask questions? Did you just listen and copy down answers? In what ways are you a good group member? How could you do better? Did your group work well? Why?)

      e)   What did you learn in this chapter? Explain how your learning relates to your response in part (d).

EF-128. **PORTFOLIO GROWTH OVER TIME - PROBLEM #1**
This growth-over-time problem will be repeated in Chapters 1, 4, and 6. So you will be doing the same problem three times, but each time you should have something new to add. After completing it for the third time you will be asked to reflect on your three responses, compare them, and discuss what you have learned.

On a separate sheet of paper (you will be handing this problem in separately or putting it into your portfolio) explain everything that you know about:

$$y = x^2 - 4 \quad \text{and} \quad y = \sqrt{(x + 4)}.$$

EF-129. **SELF EVALUATION**

In this chapter there were a number of problems in which you were expected to:
- solve linear and quadratic equations.
- write the equation of a line using slope and **y**-intercept.
- find mathematical models to fit real data.
- draw graphs of lines and parabolas.
- use the Pythagorean Theorem.
- set up ratios to solve problems.

Write a realistic self-evaluation related to the skills in the list above. Which ones do you feel confident in performing? Which skills do you still need to work on? Decide if any of the ideas listed above need to be included on your Tool Kit.

One of your responsibilities in this course will be to seek extra practice or assistance outside of the class to help you increase your confidence level. You can accomplish this task by exchanging phone numbers with other students in the class or talk with your teacher about extra help resources.

EF-130. Find the slope and intercepts of $3x + 4y = 36$. Sketch a graph.

EF-131. Write the equation of the line in the form **ax + by = c**. It will probably be easier to write the equation in $y = mx + b$ form first.

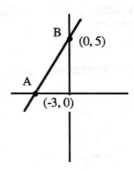

B
(0, 5)

A
(-3, 0)

EF-132. Find the length of $\overline{AB}$ in the preceding problem.

EF-133.   Suppose you have a 3 x 3 x 3 cube. It is painted on all six faces and then cut apart into 27 pieces each a 1 x 1 x 1 cube. If one of the cubes is chosen at random, what is the probability that:

a)   three sides are painted?   b)   two sides are painted?

c)   one side is painted?   d)   no sides are painted?

EF-134.   What does it mean for a shape to have a **line of symmetry**? Include several diagrams in your explanation. This might be a characteristic worth mentioning when investigating a function. Decide with your class whether or not it should be included.

EF-135.   Ayla and Sean are about to paint the side of their house facing their neighbor Brutus. Brutus tells them that since the wall they will paint on their house is only six feet away from Brutus' wall, they might as well paint his wall as well. Both walls are ten feet high, and Ayla and Sean's ladder is ten feet long. If they place the ladder exactly half way between the two buildings, and let it lean against one side and then the other (as if it were hinged on the ground), will they be able to reach the tops of both walls? Explain your answer.

EF-136.   Given: $f(x) = -\frac{2}{3}x + 3$ and $g(x) = 2x^2 - 5$,

a)   find: $f(3)$ b)   solve: $f(x) = -5$

c)   find: $g(-3)$ d)   solve: $g(x) = 9$ $\approx$ **2.65**

e)   solve: $g(x) = 8$ $\approx$ **2.55**   f)   solve: $g(x) = -7$

EF-137.   Graph the following equations.

a) $y - 2x = 3$ b) $y - 3 = x^2$

c)   State the **x** and **y** intercepts for each equation.

d)   Where do the two graphs cross? Show how to find these two points without the graphs.

EF-138.   State the domain and range of each graph.

a)   b)

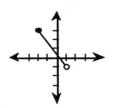   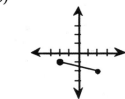

EF-139.   Think about your solutions to the previous problem. If you know a specific domain and range, what do you know about the shape of the function?

EF- 140.   **What is a PORTFOLIO?**

If you heard someone talking about an artist showing his portfolio of work, or an architect displaying her portfolio of designs, you would probably know exactly what each of those portfolios are. Those portfolios represent the artist's and the architect's best work. It is their way to demonstrate to prospective clients that they are skilled and know their trade, and that they are talented and creative in their fields.

A landscape architect might include photographs of work in progress so that he could show the stages of the job he performs. This would give the client an idea of how hard and meticulously the landscape architect works.

Sometimes this portfolio, this record of a person's work, also shows changes in the person's work. Maybe a product designer developed a new technique. She might include pictures of how her products have improved and developed over time.

Keeping all these examples in mind, what would you put in your Mathematics Portfolio? You want to make sure your best work goes into it, but this does not mean just your "neatest" or "prettiest" work. It must show that you are knowledgeable in mathematics. You must convince the person who reads your portfolio that you are competent, and in fact, good at mathematics! You must not assume, however, that the reader **knows** a lot of mathematics. Sometimes the reader will be your mathematics instructor, who does know a lot of mathematics, but you may also be showing your portfolio to other students and other instructors, some from mathematics who are not familiar with this course and others from other disciplines. So you will need to show examples of the steps you went through to produce a finished piece.

Possibly, you would include examples of work that show how you have improved or have learned new mathematics. Samples of an initial attempt followed by several revisions are important in demonstrating your ability to persevere, follow through, and grow in your understanding.

One of the reasons an Intermediate Algebra course is required, in addition to providing needed preparation in mathematics, is to determine which students are willing and able to keep going when the going gets tough. Are you a student who is willing to persevere in working on an idea or problem that is difficult or that does not make sense at first? Are you able to try a variety of approaches? Are you willing to revise? These are all skills that can be demonstrated in a portfolio.

So this year, one of your goals is to produce a Mathematics Portfolio that will be an honest reflection of the mathematics you know, have learned, and are still considering. It should be something you would be proud to show a future admissions panel or employer.

ADDITIONAL CALCULATOR PRACTICE

EFX-1. Use the graphing calculator to find the intercepts of $y = x^2 + x - 7$ accurate to two decimal places.

EFX-2. Find the coordinates of the vertex in the previous problem.

EFX-3. Find the coordinates of the intersection of $y = 2x - 5$ and $y = x^2 - 5x$ accurate to two decimal places.

EFX-4. Find the coordinates of the intersection of $y = -x + 3$ and $y = 0.5x^2 + 4x$ accurate to two decimal places.

EFX-5. Use the graphing calculator to find the intercepts of $y = \frac{1}{7}x^2 + x - 17$ accurate to two decimal places.

EFX-6. Find reasonable values to set your viewing window for the graph $y = x^2 - 16x + 44$ so that the intercepts and the vertex can be seen. Find the coordinates of the vertex.

EFX-7. Find the coordinates of the intersection of $y = 5x - 8$ and $y = \frac{1}{3}x^2 - 4x - 3$ accurate to two decimal places.

EFX-8. Find the intercepts of $y = -0.1x^2 + 5x + 100$.

LINEAR REGRESSION
Finding the Equation of the curve of best fit
(TI-81)

Note: ☐ = Key or Menu selection.

—> —> = use arrow keys to choose menu.

In numbered menus you may either press the number indicated or use arrow keys and press
ENTER

---

1. Before you begin:

Clear the regular screen.
 Clear the Y = screen.
 Set MODE
  Float 0 1 2 3 . . .
 Select 2 hit ENTER

2. To adjust Viewing Screen:
 WINDOW

 Set XMin and XMax to fit your domain.
 Set YMin and YMax to fit your range.

3. To Clear Data:

First clear previously stored data.
 2nd MATRX [STAT]
 —> —> DATA ENTER
 Select 2: ClrStat

The screen should now read: ClrStat.
 ENTER
Now you will see the regular screen with Done.

4. To Enter the Data:
 STAT
 Select 1: Edit
This will show you x1=, y1=, etc. and you can enter data points as ordered pairs.

Example: (1,1) (3,2) (5,3) (7,5) (9,8)

---

5. To See the Scatterplot:
 2nd MATRX (STAT)
 —> to DRAW ENTER
 Select 2: Scatter
The screen will show Scatter.
 Hit ENTER

6. To Calculate the Equation:
 2nd MATRX (STAT)
 —> to CALC ENTER

The screen wil show LinReg.

a is slope, b is y-intercept.
The closer the value r is to ±1, the better the fit of the line.

7. To See the Equation:

 Press Y = then VARS
 —> —> LR ENTER
 Select 4: RegEQ
 Press GRAPH

8. Now that you are done:
 Press MODE
 Select Float then ENTER

LINEAR REGRESSION
Finding the Equation of the Curve of Best Fit
(TI-82)

Note: ☐ = Key or Menu selection.

—> —> = use arrow keys to choose menu.

In numbered menus either press the number indicated or use arrow keys and press ENTER

---

1. Before you begin:

Clear the regular screen.
Clear the Y = screen.
Set MODE
   Float 0 1 2 3 . . .
Select 2 hit ENTER

---

2. To adjust Viewing Screen:
WINDOW

Set **XMin** and **XMax** to fit your domain.

Set **YMin** and **YMax** to fit your range.

---

3. To Clear Data:

First clear previously stored data.
   STAT [Statistics Key]
Select 4:ClrLst
Type 2nd 1 comma 2nd 2
The screen should now read:
ClrList L1, L2
ENTER
Now you will see the regular screen with
Done.

---

4. To Enter the Data:
STAT

Select 1: Edit
This will show you the empty data
columns. Enter all the independent
variable values in **L1.** Press
ENTER after each one. Then,
enter the dependent values in **L2.**

Example: (1,1) (3,2) (5,3) (7,5) (9,8)

---

5. To See the Scatterplot:
Press 2nd Y= for **STAT PLOT** menu

Select 1: Plot 1 ENTER

Select On ENTER

Select 1st type of graph ENTER

Select square marks ENTER

Press GRAPH

---

6. To Calculate the Equation:
   Press STAT —> CALC
   Select 5: LinReg

a is slope, b is y-intercept.
The closer the value r is to ±1, the better the fit
of the line.

---

7. To See the Equation:

Press Y = then VARS

Select 5: Statistics

—> —> EQ

Select 7: RegEq

Press GRAPH

---

8. Now that you are done:
Press MODE

Select Float then ENTER

Press 2nd Y= (STAT PLOT)

Select 4: PlotsOff

LINEAR REGRESSION
Finding the Equation of the Curve of Best Fit
(CASIO fx-7700GB)

Note: ☐ = Key or Menu Selection

Remember PRE gives you the previous menu

Use SHIFT —> to get a (,).

---

1   Before you begin:

Clear screen:
SHIFT F5 (CLS) EXE

Set Display to 2 places:
SHIFT 2 (DISP) F1 2 EXE

---

5   To enter data:

Example: (1,1) (3,2) (5,3) (7,5) (9,8)

1 SHIFT —> (,) 1 F1 (DT)
3 SHIFT —> (,) 2 F1 (DT)
5 SHIFT —> (,) 3 F1 (DT)
7 SHIFT —> (,) 5 F1 (DT)
9 SHIFT —> (,) 8 F1 (DT)

---

2   To Adjust Viewing Screen:

RANGE

Set **Xmin** and **Xmax** to fit your domain.
Set **Ymin** and **Ymax** to fit your range in the Range Menu.

---

6   To Calculate Results:

Press G<->T F6 (CAL)
       G<->T F6 (REG)

For A Press F1 (A) EXE
For B Press F2 (B) EXE

Check: A = -.45 and B = .85 r = .97

Now, graph $y = A + Bx$
              $y = -.45 + .85x$

---

3   To Clear Data from Memory:

SHIFT 3 (CLR) F2 (SCL) EXE

This step is important or previously entered data will affect your results.
Note: SCL means Statistics Data CLear

---

7   Now that you are done:

Reset to COM mode–
MODE + SHIFT 2 (DISP)
F3 EXE

---

4   To Prepare to Enter Data:

Press MODE then + (REG)
Press MODE then 4 (LIN)
Press MODE MODE then 1 (STO)
Press MODE MODE then 3 (DRAW)

Your screen should look like this:
          RUN/LIN-REG
          S-data : STO
          S-graph : DRAW
          G-type : REC/CON
          angle : Deg
          display:FIX2

---

8   Final Note:

For Logarithmic Regression
Formula: $y = A + B\ln x$
follow the same procedure except choose
5 (LOG) from the mode menu

For Exponential Regression
Formula: $y = A \cdot e^{Bx}$
follow the same procedure except choose
6 (EXP) from the mode menu
For examples refer to the Casio Manual pages 107–109

LAB/INVESTIGATION WRITE-UP

1. DESCRIPTION & PURPOSE OF THE INVESTIGATION

- Briefly explain the investigation--describe what you were required to do. DO NOT go into too much detail. Someone reading it should understand the major requirements of the investigation/lab yet they wouldn't necessarily have enough information to recreate it on their own.

- What was the purpose of the investigation?

2. DATA COMPILATION

- Include all your data -- tables, graphs, drawings, etc., --in a clear, organized format with appropriate labels.

3. DATA ANALYSIS

- Include your mathematical work, and write a thorough description of what you learned from your data and from this investigation overall. Give specifics! Address the questions from the investigation that ask you to draw conclusions. Write your answers in complete sentences in paragraph form.

- Describe the mathematical concepts and ideas you used or discovered during this investigation.

4. QUESTIONS AND COMMENTS

- What questions do you have now? These can be questions you don't feel you can answer yet or extension questions that the lab/investigation raised. Come up with at least one good question. You may include constructive comments about this activity in this section.

**NOTE**: If provided to you, carefully read the grade sheet/checklist for more specific instructions about how components of this activity are weighted.

MATHEMATICS TOOL KIT

1. Quadratic Formula

2. Pythagorean Theorem

3.

4. Fraction Busters

5. Area Formulas

6. Circle Formulas

7. Slope

MATHEMATICS TOOL KIT

8. <u>Equation of a Line</u>

9. <u>Intercepts</u>

10. <u>Sketch -vs- Graph</u>

11. <u>Domain and Range</u>

12. <u>Substitution Method</u>

13. <u>Factoring</u>

14. <u>Line of Symmetry</u>

MATHEMATICS TOOL KIT

15. <u>Elimination Method</u>

16.

17.

18.

19.

20.

21.

MATHEMATICS TOOL KIT

CALCULATOR TOOL KIT

Entering Variables

Setting the Range

Trace

Zoom

Clear Screen

Two or More Graphs

Shifting the Window

CALCULATOR TOOL KIT

Contrast

Inequalities

Powers and Roots

_____:

_____:

_____:

_____:

These grids are for use under a sheet of regular binder paper in order to save using a whole sheet of graph paper when only one graph is to be drawn.

Chapter 2
The Bouncing Ball and Other Sequences

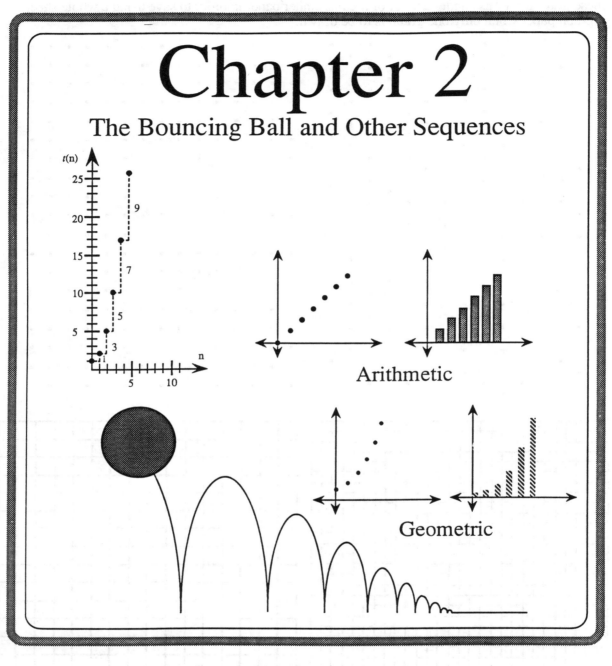

Arithmetic

Geometric

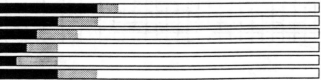

PROBLEM SOLVING
REPRESENTATION/MODELING
FUNCTIONS/GRAPHING
INTERSECTIONS/SYSTEMS
ALGORITHMS
REASONING/COMMUNICATION

CHAPTER 2

THE BOUNCING BALL AND OTHER SEQUENCES

This chapter focuses on patterns, particularly in arithmetic and geometric sequences. Throughout this chapter you will:

- continue to work in groups and use problem solving strategies as strategies for learning new mathematics.

- use patterns to make conjectures and to write algebraic representations.

- become familiar with patterns and graphs of functions that are multiplicative, or geometric, as compared to additive or arithmetic.

- determine relationships between discrete functions, such as arithmetic sequences, and continuous functions, such as straight lines.

- continue to use previously learned algebraic skills in new problem contexts, particularly in solving linear and quadratic equations, solving systems of equations and using exponents.

INTRODUCTION TO SEQUENCES

BB-1.   Lona has received a stamp collection from her grandmother. The collection is in a leather book and currently has 120 stamps. Lona joins a stamp club which sends her 12 new stamps each month. The stamp book can contain a maximum of 500 stamps.

a)   Complete the table below:

Months (n)	Total Stamps t(n)
0	120
1	132
2	144
3	
4	
5	

b)   How many stamps will she have after one year?

c)   When will the book be filled?

d)   Write an equation to represent the total number of stamps that Lona has in her collection after n months. Let the total be represented by t(n).

e)   Solve your equation for n when t(n) = 500. Explain why Lona will not exactly fill her book with no stamps remaining.

f)   Notice that we are starting with an initial value at zero months, in this chapter we will start all sequences with n = 0 as opposed to n = 1. When a sequence starts with n = 1 you have to make an adjustment in its formula.

BB-2.   Samantha was looking at one of the function machines in the last chapter and decided that she could create a sequence generating machine by connecting the output back into the input. She tried out her generator by dropping in the initial value of 8. Each output is recorded before it is recycled.

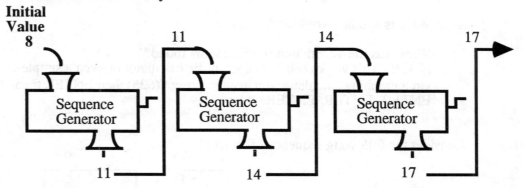

a)   The first result is 11, then 14, then 17, etc. What rule is the sequence generator using?

b)   When Samantha uses the initial value -3 and the rule "multiply by -2", what are the first five terms of the sequence?

c)   What sequence will she generate if she uses an initial value of 3 and the rule "square"?

BB-3.   Samantha has been busy creating new sequence generators and has created a bunch of sequences. Her instructor has also been busy thinking up sequences using his own devious methods. The following list (a to l) is a mixture of Samantha's sequences and her instructor's. In your groups:

    i.   For each sequence create a table like the one to the right - example (e) is shown. The initial value in the sequence we denote as t(0), the first term after the initial value is t(1), t(2) is the second term after the initial value and so on..

n	t(n)
0	2
1	3.5
2	5
3	6.5
4	8

    ii.   Continue each table for the next three terms for each sequence.

    iii.   Describe a rule for finding the next term in words or algebra.

    iv.   Decide whether the sequence could have been produced by a machine with a single simple rule for getting from one number to the next, such as repeatedly adding or multiplying by the same number.

a) 0, 2, 4, 6, 8, ...   b) 1, 2, 4, 8, ...

c) 7, 5, 3, 1, . . .   d) 0, 1, 4, 9, ...

e) 2, 3.5, 5, 6.5, ...   f) 1, 1, 2, 3, 5, ...

g) 27, 9, 3, 1, ...   h) 40, 20, 10, ...

i) -3, -1, 3, 9, ...   j) -4, -1, 2, 5, ...

k) 3, 6, 12, ...   l) 0, 1, 8, 27, 64, ...

BB-4.   Samantha thinks about another sequence generator. "If I can square, I can unsquare,"
          she reasoned, "I wonder what will happen it I start with 625 and use the rule
          unsquare." Samantha had just started her machine when the phone rang. When she
          returned her machine was generating the 4275th result after the initial value.

          a)   What is another word for "unsquare?"

          b)   What was the 4725th number Samantha found?
                 NOTE: This works better on a scientific calculator or even a simple calculator.
                 On a graphing calculator you need to press √(625) then ENTER then 2nd ,
                 ENTER, ENTER, ENTER, etc.

BB-5.   Consider the following sequence of rectangles:

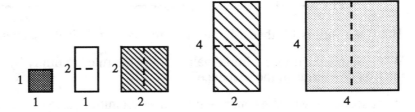

          a)   On a sheet of graph paper draw a picture of the next two rectangles in the
                 sequence of rectangles above.

          b)   Describe how each is formed from its preceding rectangle.

          c)   Write the areas of these rectangles as a sequence.

BB-6.   Write each expression below in a simpler form. The examples are reminders.

          Examples: $\dfrac{7^5}{7^3} = \dfrac{7 \cdot 7 \cdot 7 \cdot 7 \cdot 7}{7 \cdot 7 \cdot 7} = 7 \cdot 7 = 7^2$

          $(5^3)^4 = (5^3)(5^3)(5^3)(5^3) = (5 \cdot 5 \cdot 5)(5 \cdot 5 \cdot 5)(5 \cdot 5 \cdot 5)(5 \cdot 5 \cdot 5) = 5^{12}$

          a) $\dfrac{5^{723}}{5^{721}}$ b) $\dfrac{3^{300}}{3^{249}}$

          c) $\dfrac{3 \cdot 4^{1001}}{7 \cdot 4^{997}}$ d) $\dfrac{(6^{54})^{11}}{(6^{49})^{10}}$

BB-7.   Write the equation of a line with:

          a)   a slope of -2 and a y-intercept of 7.

          b)   a slope of $\dfrac{-3}{2}$ and an x-intercept of (4,0)

BB-8.   Consider the tables below. See if you can find some patterns within each table.

Table 1			Table 2			Table 3	
n	t(n)		n	s(n)		n	p(n)
0	2		0	0		0	-2
1	3		1	3		1	2
2	4		2	6		2	6
3	5		3	9		3	10

a)   How is each sequence above being generated?

b)   Extend each table a little then see if you can find a **direct** relationship between n = 5 and t(5), n = 5 and s(5), and n = 5 and p(5).

c)   Without extending the tables any further find t(25), s(25) and p(25).

d)   Write a rule that will find the nth term without going through all of the previous terms (for example table 1 would have the equation t(n) = n).

BB-9.   A triangle has vertices are A(3, 2), B(-2, 0), and C(-1, 4). What kind of triangle is it? Be sure to consider all possible triangle types. Include sufficient evidence to support your conclusion.

BB-10.   Factor each expression and simplify.

a) $\dfrac{x^2 - 4}{x^2 + 4x + 4}$ b) $\dfrac{2x^2 - 5x - 3}{4x^2 + 4x + 1}$

BB-11.   Find the x intercepts for: $y - x^2 = 6x$.

BB-12.   A dart board is in the shape of an equilateral triangle with a smaller equilateral triangle in the center made by joining the midpoints of the three edges. A dart hits the board at random. What is the probability that:

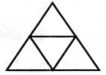

a)   the dart hits the center triangle.

b)   the dart misses the center triangle.

GRAPHING SEQUENCES

BB-13.   You may want to divide this task among the members of your group. But, be sure that each group member has his or her own set of graphs drawn. For each designated sequence in problem BB-3 draw a separate graph on the resource sheets provided (at the end of the chapter). Use **n**, the position of the term after the initial value, as the independent variable and the term itself, **t(n)**, as the dependent variable. Notice that the domain for each sequence is whole numbers. **Carefully consider whether or not to connect the points on your graphs. Discuss this with the other members of your group.**

   a)   Use the tables you made and plot the sequences from BB-3 parts (a), (b), (d), (e), (f), (i), (k) and (l).

   b)   Write the initial value and generator (the rule you found) on each graph.

BB-14.   Discuss the similarities among the graphs with your group. Write down your observations in relation to the questions below.

   a)   Which graphs look similar?

   b)   Which graphs had similar generators?

   c)   What is the significance of the initial value in each graph?

BB-15.   The graphs you drew for problem BB-13 as sequences of points might be difficult to compare and contrast . One way to learn more about the patterns is to draw "stair cases" between points and label the amounts of vertical change per one unit of horizontal change. Draw a "stair case" for each graph and write down at least two observations of the vertical and horizontal relationships.

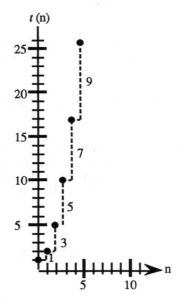

   For example, the sequence 1, 2, 5, 10, 17, . . . gives this table:

n	0	1	2	3	4
t(n)	1	2	5	10	17

   which in turn gives the graph at right.

BB-16.   Look at the sequence (c) in BB-3. Plot the sequence. How is this sequence similar to sequences (a) and (e)? Show that the sequence is similar by rewriting the generator so that it uses addition.

BB-17.   Use a similar argument to show that sequence (h) can be grouped with sequences (b) and (k). Can you find another sequence in BB-3 that would belong with this group?

♀BB-18.   **SYSTEMS OF EQUATIONS**
                          (A reminder from Elementary Algebra)

Solve each system of equations. If you remember how to do these from another
course, go ahead and solve. If you are not sure how to start, read the examples in the
following boxes first. This may be something to enter into your Tool Kit.

a) $y = 3x + 1$ b) $4x + 2y = 14$
            $x + 2y = -5$ $x - 2y = 1$

**The first subproblem in solving a system of equations is to
ELIMINATE one variable. One way to do this is by
<u>SUBSTITUTION</u>:**

Consider this system: $8y - 3x = 10$
                                    $5x + 10y = -5.$

Look for the equation that is easiest to solve for x or y. Let's try:

$$5x + 10y = -5$$
$$5x = -5 - 10y$$
$$x = -1 - 2y.$$

Now replace the x in the other equation with ($-1 - 2y$):

$$8y - 3(-1 - 2y) = 10$$
$$8y + 3 + 6y = 10$$
$$14y + 3 = 10$$
$$14y = 7$$
$$y = 0.5.$$

Find x by substituting 0.5 for y :

$$x = -1 - 2(0.5)$$
$$x = -2.$$

So the solution is (-2, 0.5).

Some people prefer to <u>ELIMINATE</u> a variable by ADDING THE TWO EQUATIONS.

First, rewrite the equations so we can see the x's and y's lined up. This makes it easier to decide what to multiply by to make the coefficients of either the x's or the y's the same.

$$-3x + 8y = 10$$
$$5x + 10y = -5.$$

We could multiply the top equation by five and the bottom equation by three to get:

$$-3x + 8y = 10 \quad \text{by 5} \rightarrow \quad -15x + 40y = 50$$
$$5x + 10y = -5 \quad \text{by 3} \rightarrow \quad \underline{15x + 30y = -15}$$

Adding: $\qquad\qquad\qquad\qquad\qquad\qquad\qquad\qquad 70y = 35$
$$y = 0.5.$$

Go back now and substitute 0.5 for y:

$$8(0.5) - 3x = 10$$
$$4 - 3x = 10$$
$$-3x = 6$$
$$x = -2.$$

Again we find the solution: (-2, 0.5). Be sure to make note of this method in your Tool Kit.

BB-19.   One week Charles bought five cans of chicken noodle soup and three cans of tuna and spent $8.12. The next week the prices were the same and his total bill was $7.52 for six cans of chicken noodle soup and two cans of tuna. Write two equations and solve them to figure out the price of a can of chicken noodle soup and the price of a can of tuna.

BB-20.   Determine whether the points A(3, 5), B(-2, 6), and C(-5,7) are on the same line. Justify your conclusion algebraically.

BB-21.   Sketch the graphs of $y = x^2$, $y = 2x^2$, and $y = \frac{1}{2}x^2$ on the same set of axes. Describe the differences between the graphs.

BB-22.   Baljit wanted to examine these equations on her graphing calculator. Rewrite each of the equations so that she can enter them into the calculator in the " y = " format.

a) $5 - (y - 2) = 3x$ b) $5(x + y) = -2$

BB-23.   At Safeway they have a special on bread and soup. Brian bought four cans of soup and three loaves of bread for $4.36. Ronda bought eight cans of soup and one loaf of bread for $3.32.

a)   Write equations for both Brian's purchase and Ronda's purchase.

b)   Solve the system to find the price for a can of soup and the price for one loaf of bread.

BB-24   Find the domain and range for each of the following functions.

a)   b)   c)   d)

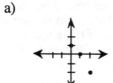   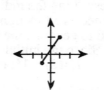   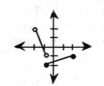

BB-25.   Simplify each of the following.

a) $\dfrac{3^5 \cdot 9^4}{27^4}$

b) $\dfrac{(b^2)^3}{(b^3 b^4)}$

c) $\dfrac{(b^n)^2}{b^n \cdot b^{n+1}}$

BB-26.   If these two cans hold the same amount of tuna, find h. Reminder: $V = \pi r^2 h$.

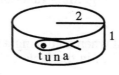

BB-27.   Toss three coins in the air. Make a list of all the possible outcomes, and find the probability that:

a)   all of them land heads up;

b)   two of them land with tails up.

CLASSIFYING SEQUENCES

BB-28.   Some sequences are named because of their common characteristics.

a)   Sequences such as (a), (c), (e) and (j) in BB-3 are called **ARITHMETIC SEQUENCES.** Describe what these sequences have in common, and give two more examples.

b)   What do the "staircases" of the arithmetic sequences have in common?

c)   Sequences such as (b), (g), (h) and (k) in BB-3 are called **GEOMETRIC SEQUENCES.** Describe what these sequences have in common and give two more examples.

d)   What do the "staircases" of the geometric sequences have in common?

e)   Sequences such as (d) and (i) in BB-3 can be called quadratic sequences. Why? Concentrate on sequence (d).

f)   How do these quadratic "staircases" compare with those of the geometric sequences?

g)   The sequence in (f) from BB-3 is special and is named after the mathematician who discovered, explored, and wrote about it. His name was Fibonacci. What is different about the staircase of this sequence?

♀BB-29.   **Notation for writing sequences.** When we write a sequence such as 3, 7, 11, 15, ... , we need to have a way of describing the individual terms. The most common way to do so is to think of the sequence as a function whose domain is the non-negative integers. The expression t(n) is used to generate the term, t, of the sequence as a function of the position number, n. Since the initial value is 3, t(0) = 3. We could write this sequence as: t(0) = 3; t(1) = 7; t(2) = 11; etc.

a)   Find t(3), t(4), and t(5). A table might help.

b)   Find t(25). Try to use the relationship between 3 and t(3), 4 and t(4), 5 and t(5) to figure this out so you don't have to make a huge table.

c)   Find t(n).

d)   Find t(100).

e)   Put information regarding function notation for sequences in your Tool Kit.

BB-30.   A table is very useful when analyzing a sequence to determine its pattern. Consider the sequence 7, 10, 13, . . . in the table below:

n	0	1	2	3	4	5	6
t(n)	7	10	13	16			

a)   Fill in the rest of the table.

b)   What is the initial value and the generator for the sequence? What kind of sequence is it?

c)   The generator in an arithmetic sequence is frequently referred to as the common difference. Why might this term be used?

d)   How many 3's had to be added to the initial term to get the term for n = 4?

e)   What is an algebraic expression for the value of the nth term?

f)   graph this function and find the slope of the line which would go through the points on the graph.

BB-31.   In the sequence of the last problem, is it possible for 30 to show up in the t(n) row? Explain your thinking.

BB-32.   What kinds of sequences have a graph whose points lie on a line? What specific information will you know about the line based on the nature of the sequence?

BB-33.   Using what you have learned in the previous problems find the slope of the line containing the sequence of points listed below. Write an expression for the nth term in each case.

a)   5, 8, 11, 14, . . .   b)   3, 9, 15, . . .

c)   26, 21, 16, . . .   d)   7, 8.5, 10, . . .

♀ BB-34.   Find the function t(n) where the initial value is "a" and the generator is "adding d" (the common difference or rate of change).

BB-35.   Find the initial value and the common difference for a sequence where t(5) = -3 and t(50) = 357.

One (not the only) way to do this is to use the equation you found in the previous problem and substitute -3 for t(n) and 5 for n. Then use the equation again and substitute 357 for t(n) and 50 for n. Now review the procedures in BB-18 and solve the two equations you wrote to find **a** and **d**.

BB-36.   Write a paragraph about the relationship between an arithmetic sequence and its graph. Be sure to explain the relationships among the common difference, slope, y-intercept, and the initial value of the sequence.

BB-37.   In the sequence 1, 4, 7, 10, 13, 16,..., we know t(2), = 7 and t(5) = 16.

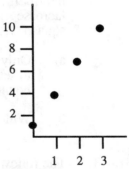

a)   What is t(8)?

b)   What is t(14)?

c)   The terms, t(0), t(1), t(2), t(3), of this sequence are graphed for you at right.

d)   What is the equation of the line that passes through these points? Write your equation in the form t(n) =

e)   What is the significance of the initial value in terms of the graph?

f)   What does the slope of the line represent?

BB-38.   Solve the system below for m and b:

$$15 = 5m + b$$
$$7 = 3m + b$$

BB-39.   The previous problem could have been created by substituting the points (5, 15) and (3, 7) into the equation y = mx + b. For the first equation we replace x with 5 and y with 15. We get the second equation by replacing x with 3 and y with 7. By solving the system, you can find the equation of the line through the two points. What line is determined by these two points?

BB-40.   Using what you learned in the previous two problems, find the equation of the line through the points (2, 3) and (5, -6). The first equation you need is 3 = 2m + b.

BB-41.   Given that n represents the length of the bottom edge of the L-shaped figures below, and t(n) is the total number of dots in the L-shape, what numbers are generated? Find t(46). What is t(n)?

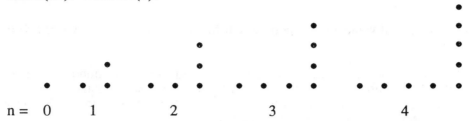

n =   0   1   2   3   4

BB-42.   Many games depend upon the players' response to the bounce of a ball. For this reason manufacturers have to make balls so that their bounce conforms with certain standards. Some published standards for different balls are:

tennis ball... rebound 106 cm to 116 cm when dropped from 200 cm
handball..... rebound 62" to 65" when dropped from 100" at 68° F
lacrosse ..... rebound 45" to 49" when dropped from 72" on a wooden floor
squash ...... rebound 28" to 31" when dropped from 100" onto a steel plate at 70° F

a)   Only squash authorities recognize the importance of both temperature and surface in setting their standard. What effect could surface and temperature have on the bounce of a ball?

b)   Which ball bounces proportionately higher than the others?

BB-43.   The following data points show the actual score and the adjusted score for a recent Algebra 2 quiz. Graph these data points and write the equation used by the teacher to adjust the scores.

score	27	15	20	18
adjusted	91	55	70	?

a)   What score would have been adjusted to 82?

b)   What is the adjusted score for 18?

c)   What does the slope of this line represent?

BB-44.   Two congruent overlapping squares are shown. If a point inside the figure is chosen at random, what is the probability that it will be in the shaded region?

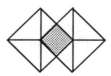

FOLLOW THE BOUNCING BALL: CHAPTER 2 LAB
(BB45 to BB-51)

BB-45.   Designate one person in your group as recorder, one as ball dropper, and at least two as spotters. Make a table like the one below. Choose a starting height and record it in your table. Predict what you will observe when the ball is dropped from this height and rebounds. Drop the ball from the chosen height and record the rebound height of the ball. **Make at least three more trials from this starting height** and record each resulting rebound height. It will be easier to compare heights if you measure to the **bottom** of the ball in both cases.

a)   What do you notice about the rebound heights?

b)   Select another 5 starting heights and repeat the procedure for each one (so you will have at least fifteen entries in your table).

Remember to measure both heights to the bottom of the ball.

starting height	rebound height	rebound ratio

BB-46.   Identify the independent variable as **x** and the dependent as **y** and graph your results carefully on graph paper. What do you notice? Draw a **line of best fit**. Write the equation for your line.

BB-47.   For each trial, make a third column on your table and calculate the ratio of the rebound height to the starting height, r/s. What do you notice?

a)   How is this result related to the slope of the line of best fit?

b)   Explain why it makes sense for the y-intercept to be zero.

c)   Physicists call relationships such as this one between the height at which the ball was dropped and the rebound ratio <u>direct variation</u>. Why do you think they use this description? Think of another direct variation relationship, describe it and write an equation for it.

BB-48.   Use the information you have gathered to predict how high the superball will rebound the first time if dropped from two meters.

a)   How does the predicted rebound height compare to the actual rebound height? Why might they be very close? What might explain any discrepancy?

b)   Suppose the superball is dropped and you notice that its rebound height is 60 cm. From what height was the ball dropped?

c)   Suppose the superball were dropped from a window 200 meters up the Empire State Building. What would you predict the rebound height to be after the first bounce? Is your prediction reasonable considering the nature of a super ball? Explain your thinking on this issue.

d)   How high would the ball rebound after the second bounce? the third?

e)   What factors might affect the accuracy of your predictions?

BB-49.   Now drop the superball from 2 meters and allow it to bounce three times.

a)   Record the rebound height after the first, second and third bounces. Repeat this experiment several times.

b)   Now drop the ball from the first rebound height and record the height of the next two bounces. Repeat this experiment several times.

c)   Compare the two sets of data. Write about the similarities and differences between the data sets.

BB-50.   Suppose we drop a superball from a height of ten feet and let it bounce. Make a graph of the rebound height for each bounce. The independent variable is the number of bounces. Your graph will show a record of the height for each bounce, not a picture of the ball bouncing.

a)   Should the points on your graph be connected? Explain your thinking.

b)   In general, how could you determine whether or not the data points on your graph should be connected? Give an example of a data set that would be connected and one that would not.

c)   When the points on a graph are connected, and it makes sense to connect them, we say the graph is **continuous**. If it is not continuous, and is just a sequence of separate points, we say the graph is **discrete**. Look back at the graph you drew in BB-49. Should the graph in BB-49 be continuous? Explain your thinking on this.

d)   Add the definitions of discrete and continuous to your Tool Kit.

BB-51.   The saga of the Bouncing Ball will continue so be sure your data is accurate, neatly organized, and handy. You will be using your data later to answer each of the following questions:

a)   What equation could represent your graph in problem BB-50?

b)   How long will it take for the ball to stop bouncing?

BB-52.   **t(n)** is often called the **nth** <u>term</u> of the sequence, for example, t(5) is term number 5 or the 5th term. For each sequence determine what number **n** could be if 447 is the **nth** term. Is it always possible for 447 to be a term?

a) $t(n) = 5n - 3$ b) $t(n) = 24 - 5n$

c) $t(n) = -6 + 3(n - 1)$ d) $t(n) = 14 - 3n$

e) $t(n) = -8 - 7(n - 1)$

BB-53.   Choose one of the sequences in the previous problem for which 447 was **not** a term. Write an explanation clear enough for an elementary algebra student to understand how you were able to determine that 447 was not a term of the sequence.

BB-54.   Seven years ago Raj found a box of old baseball cards in the garage. Since then he has added a consistent number of cards to the collection each year. He had 52 cards in the collection after 3 years and now has 108 cards. How many cards were in the original box?

a)   Raj plans to keep the collection for a long time. How many cards will the collection contain 10 years from now?

b)   Write an expression that determines the number of cards in the collection after n years. What does each number stand for?

BB-55.   Dr. Sanchez asked her class to simplify x + .6x

Terry says, x + .6x = 1.6x

Jo says, x + .6x = .7x

Whose response is correct? Justify your conclusion.

BB-56.   Recall the sequence (h) from BB-3. The sequence was 40, 20, 10, . . .

a)   What kind of a sequence is 40, 20, 10, ...? Why?

b)   Plot the sequence up to n = 6.

c)   Since the sequence is decreasing, will the values ever become negative? Explain.

BB-57.   Solve the system by graphing each line and estimating their point of intersection, then solve the system algebraically to check.

$$x + y = 5$$
$$y = \frac{1}{3}x + 1$$

BB-58.   Solve each of the following equations.

a) $\frac{m}{6} = \frac{15}{18}$ b) $\frac{\pi}{7} = \frac{a}{4}$

BB-59.   Find the **x** and **y** intercepts for the graph of: $y = x^2 + 4x - 17$

BB-60.   Plot the points (3, -1), (3, 2) and (3, 4). Draw the line through these points. The equation of this line is x = 3. Why do you think the equation is defined this way?

a)   Plot the points (5, -1), (1, -1) and (-3, -1). What will be the equation of the line through these points? Explain.

b)   Choose any three points on the y-axis. What will be the equation of the line that goes through those points?

BB-61.   How did you answer BB-55? Did you agree with Terry? If so, do these problems. If not, go back and reconsider. Remember: x + .6x = 1x + .6x and this is the same as 1.0x + .6x.

a)   y + .03y   b)   z - .2z   c)   x + .002x

BB-62.   The sequence 5, 10, 15, ... has 5 as the first term, 10 as the second term, and so on. Suppose this sequence is part of another arithmetic sequence where 5 is the first term and 10 is the third. What would the new sequence be?

BB-63.   Consider the following sequences:

| **Sequence 1** | **Sequence 2** | **Sequence 3** |
| 2, 6,. . . | 24, 12, . . . | 1, 5,. . . |

a)   Assuming that the sequences are arithmetic, find the next four terms for each sequence above. For each sequence, write an explanation of what you did to get the next term or write a rule.

b)   Would your terms be different if the sequences were geometric? Find the next four terms for each sequence if they are geometric. For each sequence, write an explanation of what you did to get the next term or write a rule.

BB-64. Write the range and domain for each of the following graphs.

a) 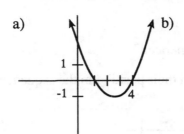 b) 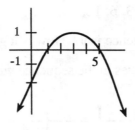 c)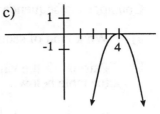

BB-65. Examine the cases below:

Case 0	Case 1	Case 2	Case 3	Case 4
1	1	1	1	1
	1 1	1 1	1 1	1 1
		1 2 1	1 2 1	1 2 1
			1 3 3 1	1 3 3 1
				1 4 6 4 1

a) What would Case 5 look like? Draw it and explain how you arrived at your conclusion.

b) This arrangement is called Pascal's Triangle and is very useful in the calculation of probabilities among other things. Describe a rule to generate any case.

BB-66. Examine each of the following sequences. For those which are arithmetic sequences, find the expression for the nth term. For those that are geometric, find the generator (the multiplier) used to get from one term to another.

a) 1, 4, 7, 10, 13, … b) 0, 5, 12, 21, 32, …

c) 2, 4, 8, 16, 32, … d) 5, 12, 19, 26, …

e) x, x+1, x+2, x+3, … f) 3, 12, 48, 192, …

BB-67. A dart hits each of these dart boards at random. What is the probability that the dart will land in the shaded area?

a) b)

GEOMETRIC SEQUENCES

♀BB-68.   Consider the sequence 3, 6, 12, 24, . . .

a)   What kind of sequence is it? How is it generated?

b)   Pairing off the values in the sequence with their position in the sequence we get
      the table below:

n	0	1	2	3	4	
t(n)	3	6	12	24		

Create your own table and continue it to include the next five terms of the sequence.

c)   How many times do we multiply the initial value of 3 by 2 in order to obtain the
      value of 24 in the sequence?

d)   Is there a short cut for doing repeated multiplication? What is it?

e)   The value found when n=6 can be obtained by multiplying $3 \cdot 2 \cdot 2 \cdot 2 \cdot 2 \cdot 2 \cdot 2$.
      Rewrite this expression using exponents

f)   What will the representation be for the hundredth term of the sequence?

g)   How could you represent t(n)?

♀BB-69.   Write a function **t(n)** where the initial value is **a** and the multiplier is **r**. (The multiplier
           is sometimes called the common ratio. That's why we chose **r**.)

BB-70.   A tank contains 8,000 liters of water. Each day, one-half of the water in the tank is
          removed. How much water will be in the tank after:

a)   the 6th day?

b)   the 12th day?

c)   the nth day?

BB-71.   **FLU EPIDEMIC HITS LOCAL AREA!!! 30% increase in cases reported
          this week.** Today there were 100 cases reported to begin the annual flu season. If
          this rate continued how many cases would be reported 4 weeks from now?

a)   Make a table for weeks zero to four.

b)   How many cases would be reported after 7 weeks?

c)   Write a general expression that enables you to determine the number of cases
      reported after n weeks.

d)   For how long do you think this pattern is likely to continue?

BB-72.   The Return of the Bouncing Ball! Refer to the data that you collected earlier in this chapter when conducting the Bouncing Ball Investigation (problem BB-50). Use that data and the graph generated from the data to determine an equation that could represent your graph.

BB-73.   How high will the ball be after the 12th bounce?

BB-74.   How long will it take for the ball to stop bouncing?

BB-75.   Sketch $y = x^2$, $y = -3x^2$, and $y = -0.25x^2$ on the same set of axes. What does a negative coefficient do to the graph?

BB-76.   Look back at the data given in BB-42 regarding the rebound ratio for an approved tennis ball. Select a rebound ratio from within the stated tennis ball range. If you dropped your tennis ball from a height of 10 feet:

a)   how high would it bounce on the first bounce?

b)   how high would it bounce on the tenth bounce?

c)   how high would it bounce on the nth bounce?

BB-77.   Solve: $\dfrac{2 + x}{3} - \dfrac{x - 1}{2} = \dfrac{x}{4}$

BB-78.   Find the values of the sequence for n = 0 to n = 4.

a) $t(n) = 8 + 7(n - 1)$ b) $t(n) = -5$

c) $t(n) = 2^{n-4}$ d) $t(n) = (-2)^n$

e)   What is the domain for the sequences?

f)   What are the ranges for the sequences?

BB-79.   A sequence is defined as $t(n) = 4n - 3$. List the first four terms.

a)   What type of sequence is this? Explain. Could you have answered this question without finding the first four terms? Explain.

b)   321 is in this sequence. What term number is it? Explain how you found this answer.

BB-80.   **ClothTime Problem:** What a deal! ClothTime is having a sale: 20 percent off. Beth cruises over and buys 14 cool shirts. When the clerk rings up her purchases, Beth sees that the clerk has added the 5 percent sales tax first, before taking the discount. She wonders whether she got a good deal at the store's expense or whether she should complain to Ralph Nader about being ripped off. Your mission is to find out which is better: discount first, then tax, or tax first, then discount. Before you begin calculating, make a guess.

♀BB-81.   In the preceding problem, suppose that the total for the shirts Beth was buying was "x" dollars.

a)   What is the tax (in terms of x)?

b)   What is the total cost in terms of x?

c)   The coefficient in (b) is what we call a **multiplier.** It is greater than one because the tax made the cost increase. What would be the multiplier for a tax rate of 7%?

d)   Convert each of these increases to a multiplier:

3 percent   8.25 percent   2.08 percent

e)   Record the term **multiplier** in your Tool Kit along with an understandable definition or explanation.

BB-82.   RECTANGULAR NUMBERS

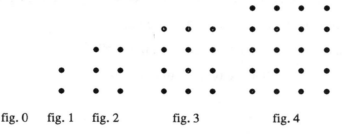

fig. 0   fig. 1   fig. 2   fig. 3   fig. 4

The sequence above has the terms 0, 2, 6, 12, 20 . . . where the value of the sequence is the number of dots in the figure.

a)   What is the 10th rectangular number?

b)   What is the 100th rectangular number?

c)   What is the nth rectangular number?

BB-83.   In 1992, Charlie received the family heirloom marble collection, consisting of 1239 marbles. The original marble collection was started by Charlie's great grandfather back in 1898. Each year Charlie's great grandfather had added the same number of marbles to his collection. When he passed them on to his son he insisted that each future generation add the same number of marbles per year to the collection. When Charlie's father received the collection in 1959 there were 810 marbles.

a)   How many marbles are added to the collection each year?

b)   How many years has the collection been maintained?

c)   Use the information you found in part (b) to figure out how many marbles were in the original collection when Charlie's great grandfather began it.

d)   Write a generalized expression describing the growth of the marble collection since it was started by Charlie's great grandfather.

e)   When will Charlie have more than 2000 marbles?

BB-84.   In a geometric sequence t(0) is 3 and t(2) is 12. Find t(1) and t(3). Write a sentence to explain how you could start with the first term, 3, and generate the other terms of this sequence.

MULTIPLIERS AND APPLICATIONS

BB-85.   Karen works for Macy's and receives a 20% discount on any purchases that she makes. Today Macy's is having their end of the year clearance sale where any clearance item will be marked 30% off. When Karen includes her employee discount with the sale discount, what is total discount she will receive? Does it matter what discount she takes first?

Using graph paper, separate two 10 by 10 grids as shown below.

CASE 1 - 20% discount first   **CASE 2** - 30% discount first
　　　30% discount second   　20% discount second

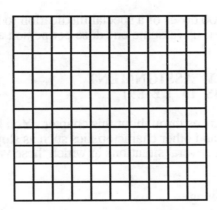 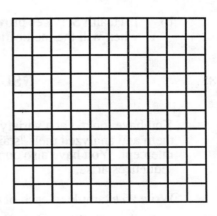

a)   Each grid has 100 squares - take the first discount in each case by shading the appropriate number of squares.

   i)   How many squares remain after the first discount in case 1?

   ii) How many squares remain after the first discount in case 2?

b)   Using the remaining squares in each case take the second discount - Note: You no longer have 100 squares to start.

   i)   How many squares remain now in case 1?

   ii) How many squares remain now in case 2?

c)   If you start by taking a 20% discount, what percent remains?

d)   If you start by taking a 30% discount, what percent remains?

e)   Multiply these two remaining percents together - what percent do you now have? How does these relate to Karen's problem above?

BB-86.   Starting with x in each case, find a simplified form for each increase or decrease. For example: A 30% increase is represented: $x + 0.3x = 1.3x$

    a)   A 5% increase   b)   A 12% discount

    c)   An 8.25% tax   d)   A reduction of 22.5%

BB-87.   Remember the flu epidemic? Well it has become a statewide crisis, but you have also developed a deeper understanding of how a multiplier can help you solve a problem such as this. If Los Angeles had 1250 cases to start their epidemic, with an increase of 27% per week, how many cases would be reported in L.A. after 4 weeks of the epidemic?

    a)   Write the general expression from which you can determine the number of cases in any week of the L.A. flu season.

    b)   If Health Services provided 8,000 doses of a special medication to fight the virus, when will the medication be used up (assume one dose per person affected)?

BB-88.
SUMMARY ASSIGNMENT
(BB-88 to BB-89)

You have looked at a variety of sequences throughout this chapter. You have graphed them and analyzed their "stair cases" and other characteristics such as slope and y-intercept. You have also come across several situations in which these various sequences arise.

Write an example of each of the following:
    a)   An increasing arithmetic sequence.
    b)   A decreasing arithmetic sequence.
    c)   An increasing geometric sequence.
    d)   A decreasing geometric sequence.

Each example should include
    1)   The sequence.
    2)   A graph of the sequence.
    3)   An equation for the sequence.
    4)   A description of what the various parts of the sequence represent.
    5)   A description of how the parts of the graph relate to its sequence.

BB-89.   Write a problem that could be solved by creating and analyzing a sequence. The sequence could be arithmetic, geometric, or a combination of both. Be sure to solve your problem. You will be sharing these with members of your group who will help you to revise your problem for your portfolio.

BB-90.   **GROWTH OVER TIME - PROBLEM #2**

On a separate piece of paper (so you can hand it in separately or add it to your portfolio) explain **everything** that you now know about:

$$f(x) = 2^x - 3.$$

BB-91.   John finished his homework last night, but his brother colored all over the paper. All that remained was:

5, 12, 19,   390, 397

He knew that 5 was the initial term and 397 the last, but he did not remember anything else about the sequence.

a)   What is the 40th term?

b)   How many terms after the initial value is the last value in John's sequence?

BB-92.   The multiplier principle can be used if there is a percent decrease. Suppose there is a 20 percent discount. Again let **x** be the original cost.

a)   What is the discount (in terms of **x**) ?

b)   What is the new price in terms of **x**?

c)   The coefficient in (b) is a **multiplier**. It is less than one because the discount made the cost decrease. What would be the multiplier for a sale rate of 15% off?

BB-93.   Find the multiplier that is determined by a

a)   3 percent decrease   b)   25 percent decrease   c)   7.5 percent decrease

BB-94.   Use the idea of a multiplier to look back at the ClothTime problem and explain why both answers were the same.

BB-95.   Find the value of y in the sequence 2, 8, 3y + 5, . . .

a)   if the sequence is arithmetic.

b)   if the sequence is geometric.

BB-96.   Consider the following tables:

n	t(n)
0	100
1	50
2	
3	

x	f(x)
0	50
1	
2	100
3	

a)   Copy and complete each table to make the sequences arithmetic.

b)   Copy and complete each table to make the sequences geometric.

c)   Write the rule for each sequence in (a) and (b) above.

BB-97.   Graph $y = x^2 + 3$ and $y = (x + 3)^2$. What are the similarities and differences between the graphs? How would these graphs compare to $y = x^2$?

BB-98.   **TRIANGULAR NUMBERS**

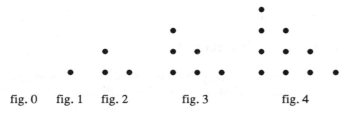

fig. 0   fig. 1   fig. 2   fig. 3   fig. 4

The sequence above has the terms 0, 1, 3, 6, 10. . . where the value of the sequence is the number of dots in the figure. What problem from a few days ago would be related to this sequence? Find it. It will be very useful in figuring out the answers.

a)   What is the 15th triangular number?

b)   What is the 200th triangular number?

c)   What is the nth triangular number?

BB-99.   **SELF EVALUATION**

You did a self evaluation of your skills at the end of Chapter 1. Here are a few more skills for you to check. Are you confident that you can:

•   solve a system of equations?

•   solve any quadratic equation which has real solutions?

If you are not confident in these skills, ask for help from your group or your teacher. You may need to seek some review help and do some extra practice outside of classtime. Find a sample problem that is worked out completely and correctly. Enter this sample into your Tool Kit.

EXTRA PROBLEMS

BB-100.   You decide to hire two junior high school students to rake the leaves in your yard, rather than do it yourself. They will work a couple hours each afternoon until the job is completed. (You assume that this will take no more than three weeks). You give them a choice about which payment plan they prefer. Plan A pays $11.50 per afternoon, while Plan B pays 2 cents for one day's work, 4 cents for 2 days' work, 8 cents the three days' work, 16 cents for four days, and so on. Each student chooses a different plan. On which day will they be paid approximately the same? Support your thinking with data charts and/or graphs.

BB-101.   Your favorite radio station, KWNS is having a contest. The D.J. poses a question to the listeners. If the caller answers correctly, he/she wins the money in the pot. If the caller answers incorrectly, $20 is added to the pot and the next caller is eligible to win. There is a major stumper question that no one has answered correctly and the pot is full of money!!

a)   You were the 15th caller today and you won $735!! How much money was in the pot at the beginning of today? Careful, you need to think about how many times the pot was increased today.

b)   Suppose the contest starts with $100. How many people would have to guess wrong for the winner to get $1360?

BB-102.   The spiral below is made using a sequence of right triangles, each with a leg that measures one unit, and the second leg being the hypotenuse of the triangle before it. Although only four right triangles are shown, we could continue this spiral forever. As you work on this problem, **think back to previous function investigations that you have conducted. Be thorough and provide relevant data, graphs, and conclusions.** Here you will be investigating two functions. Each is described below:

i.   Consider a function where the input is the number of right triangles in the figure, and the output is the length of the hypotenuse of the last triangle.

ii.   Now, consider the function where the input is the number of right triangles in the figure, and the output is the area of the last triangle formed.

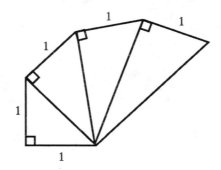

BB-103.   The first three terms of a sequence are **x, y, z**.

   a)   If the sequence is arithmetic.

       i) What do we know about y - x and z - y?

       ii) Use this to solve for **z** in terms of **x** and **y**.

   b)   If the sequence is geometric.

       i) What ratio is equal to $\frac{y}{x}$?

       ii) Solve for **z** in terms of **y** and **x**

BB-104.   A rookie basketball player was recently drafted by the Phoenix Suns. His salary will be $673,500 for the first year, with an increase of 20% each year of his five year contract. How much will he make in the fifth year of his contract?

   a)   A teammate is currently in the fourth year of a five year contract. His current salary is $875,900, and this represents an annual raise of 15% for the life of his contract. What was his starting salary?

   b)   Write the expression that describes the terms of the contract for each player. When you think you have the equations, be sure to check your answers to see if they work.

   c)   What was the total amount of each five year contract for each of these players?

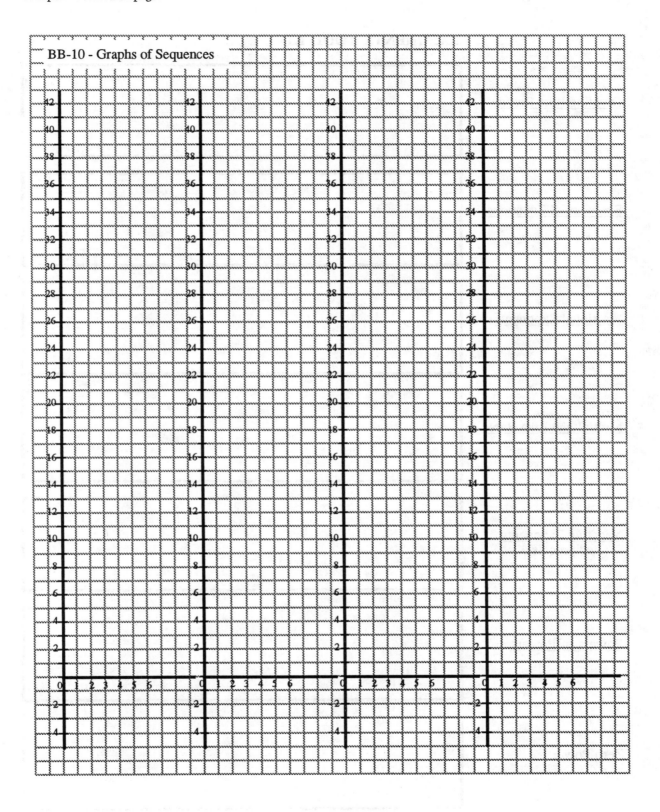

BB-10 - Graphs of Sequences

EXPONENTS TOOL KIT

Dividing Like Bases

Exponential Equations

Fractional Exponents

Multiplying Like Bases

Negative Exponents
(Reciprocals)

Powers of Powers

Zero Power

Chapter 3

Exponential Functions
(Fast Cars)

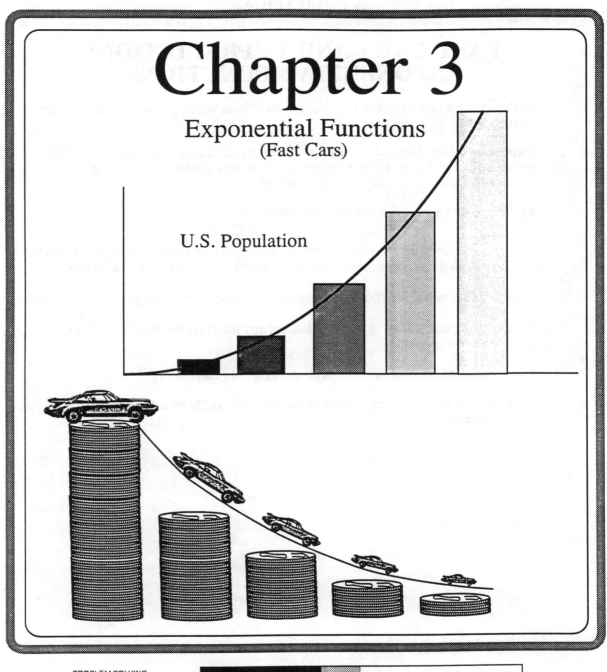

U.S. Population

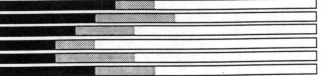

CHAPTER 3

FAST CARS AND DEPRECIATION: EXPONENTIAL FUNCTIONS

In this chapter, we examine exponential functions starting from a basis of geometric sequences.

Exponential functions can be used to represent situations that involve growth or decay, appreciation or depreciation, so they are often very useful as mathematical models in science, business, banking, and economics.

In this chapter you will have the opportunity to:

- see a relationship between geometric sequences and exponential functions that is similar to the relationship between arithmetic sequences and linear functions.

- use exponential functions to represent situations modeling growth and decay.

- develop your ability to write equations and interpret the meaning of fractional and negative exponents.

- continue to become familiar with the graphing calculator.

- continue to practice the use of basic algebraic skills in solving some not so basic problems.

The following problem is the theme problem for this chapter. While it might be possible to solve it today, it will be easier to solve it later in the chapter when it will re-appear. For now read it and go on to problem FX-2.

FX-1.   **Fast Cars:** As soon as you drive a new car off the dealer's lot, the car is worth less than what you paid for it. This is called **depreciation.** Chances are that you will sell it for less than the price that you paid for it. Some cars depreciate more than others (that is, at different rates), but most cars depreciate. On the other hand, some older cars actually increase in value. This is called **appreciation.** Let's suppose you have a choice between buying a 1993 Mazda Miata for $17,000 which depreciates at 22% a year, or a used 1990 Honda CRX for $12,000 which only depreciates at 18% a year. In how many years will their values be the same? Should you instead buy a 1967 Ford Mustang for $4,000 that is appreciating at 10% per year? Which car will have the greatest value in 4 years? In 5 years?

MULTIPLIERS AND COMPOUND INTEREST

FX-2.   Spud's uncle, Nathan, offers to pay him an 8 percent annual interest rate to encourage him to save some money for his education. Unlike a bank, Uncle Nathan will pay simple interest. Banks generally pay compound interest, but we want to start by looking at simple interest. Suppose Spud deposits $100 with his uncle and Nathan pays interest on the $100 at the end of each year. The "zero" year is the initial deposit.

SIMPLE INTEREST

year	amount of money, in dollars
0	100.00 (initial value)
1	108.00
2	116.00
3	124.00
4	132.00

a)   By what percent had Spud's account increased at the end of the fourth year?

b)   Continue this table for the next eight years.

c)   Draw a graph to show the relationship between years (independent variable) and amount of money in the account. Label each axis **and** state whether the graph represents an arithmetic sequence, a geometric sequence, or neither.

FX-3.   As you may know, banks and credit unions do not pay interest in this simple way. They normally pay compound interest. That means that interest is paid not only on the amount invested (called the **principal** or **initial value**) but **also** on the accumulated interest. In other words, you get interest on your interest. To see this clearly, look at the table for calculating compound interest. We are using the same interest rate, eight percent per year, and the principal is still $100.00.

COMPOUND INTEREST

year	amount of money, in dollars
0	100.00 (initial value)
1	108.00
2	116.64
3	125.97
4	

   a)   Explain to other members of your group why the amount in the account at the end of the second year would be $116.64.

   b)   Copy the table and enter the appropriate values for the 4th year. Be sure to round off appropriately (you can't have fractions of a cent).

   c)   The simple interest that Spud's uncle paid increased his account by a total of 32% over the four years. Based on the compound interest table above, by what percent would the account increase?

FX-4.   a)   Extend the table of compound interest for another eight years.

   b)   Draw graph for the table of compound interest. Label each axis **and** state whether the graph represents an arithmetic (linear) sequence, a geometric sequence, or neither.

   c)   Are the graphs for the last two problems **discrete** (points only) or **continuous** (connected)? Explain.

FX-5.   Look back at the table you just made for compound interest.

   a)   Explain how to get an amount in any line from the line above it.

   b)   Represent the amount of money in your account after two, three, and four years using powers of 1.08.

   c)   Suppose you invested $1000 in this credit union and left it there for twenty years. How much money would be in your account then?

FX-6. Suppose you have a young cousin and yesterday was her sixth birthday. She got $100 from your grandfather and you want to convince her to open a savings account with that $100 at the credit union. In order to convince your cousin, you need to explain to her how much money she will have by her eighteenth birthday if the credit union is paying at an 8% annual interest rate compounded <u>quarterly.</u>

    a) If 8% is the annual interest rate, what is the quarterly interest rate? What is the multiplier?

    b) Write an expression that represents the amount of money in the account after 10 quarters.

    c) Write an expression that represents the amount of money in the account after x quarters.

    d) Write an expression that represents the amount of money in the account on your sister's 18th birthday **and** find the value of that expression.

FX-7. What if you could earn 12% per year (3% per quarter) and you started with $356. Use the expression you wrote in the last problem as a model and write an equation for this new function where x still represents the number of quarters and y represents the amount of money you end up with.

FX-8. The concert has been sold out for weeks, and as the date for the concert draws closer, the price of the tickets increases. The cost of a pair of concert tickets was $150 yesterday and today it is $162. Assuming that the cost continues to increase at this rate:

    a) What is the daily rate of increase What is the multiplier?

    b) What will be the cost one week from now (the day before the concert)?

    c) What was the cost two weeks ago?

FX-9. Solve each of the following for x.

    a) $2^3 = 2^x$ b) $x^3 = 5^3$

    c) $3^4 = 3^{2x}$ d) $2^7 = 2^{2x+1}$

FX-10. Mindy, an only child, is constructing a family tree and notices a pattern. She notices that she has two first-generation predecessors (her natural parents), four second-generation (her grandparents), eight third-generation, and so on. Create a table to show the number of natural parents, grandparents, great grandparents, etc. as a function of the generation number.

    a) If Mindy's grandparents are the second generation and her parents are the first generation, which generation is she?

    b) She had $8 = 2^3$ great-grandparents, $4 = 2^2$ grandparents, and 2^1 parents. The number of Mindys is the initial value, so how many Mindys are there? What power of 2 represents the number of Mindys?

FX-11.   Consider the following pattern:

$$2^4 = 16, \qquad 2^3 = 8, \qquad 2^2 = 4, \qquad 2^1 = 2.$$

a)   What should $2^0 = ?$

b)   Explain the relationship between this problem and the previous problem.

FX-12.   Start with a rectangular sheet of scratch paper.

a)   Now fold the paper in half. How many times has the paper been folded? How many rectangular regions are there?

b)   Fold it in half again. With two folds, how many regions are there?

c)   Three folds. How many regions?

d)   Four folds? Five folds? What is the pattern or rule?

e)   Write an equation to represent the rule. Let y represent the number of regions and x the number of folds.

f)   How many regions were there before you folded the paper in half the first time, in other words, when there were zero folds? How does your equation describe this situation?

Zero Power. In the previous problems you found the value of 2^0 by looking at patterns. What does your calculator give for the value of 2^0? What value(s) do you get for $3^0, 4^0, 10^0, 101^0$? Experiment with lots of different values raised to the zero power (try $(-2)^0$, $(\frac{3}{5})^0$, and at least three others). Write down your results. Try more negative numbers raised to the zero power. Try fractions and decimals. Write a general rule about numbers raised to the zero power. Should this be in your Tool Kit?

Remember you can refer to your Tool Kit for a review of how to use the quadratic formula.

FX-13.   Solve $2x^2 - 3x - 7 = 0$. Give your solutions both in radical form and as decimal approximations.

FX-14..   Write each of the following in its smallest base. (<u>Examples</u>: 16 can be written as 2^4, and $27^2 = (3^3)^2 = 3^6$.)

a)   64   b) 8^3 c) 25^x

d) 16^{x+1} e) $\dfrac{16}{81}$ f) 81^2

FX-15.   Solve the following systems of equations.

Which system is most efficiently solved by substitution? Explain.

Which system is most efficiently solved by elimination? Explain.

a) $3x - 2y = 14$
        $-2x + 2y = -10$

b) $y = 5x + 3$
        $-2x - 4y = 10$

FX-16.   Give the coordinates of the x- and y-intercepts for each of the following functions.

a) $f(x) = x^2 + 6x - 72$

b) $g(x) = -5x + 4$

FX-17.   If you flip a fair coin:

a)   what is the probability of getting a "heads?"

b)   what is the probability of getting a "tails?"

FX-18.   **THE PENNY LAB**

The Experiment: With your group, conduct the following experiment.

Trial #0:   Start with 100 pennies. This is your initial value.

Trial #1:   Dump the pennies in a pile on your desk. Remove any pennies that have "tails" side up. Record the number of pennies <u>left</u> , the "heads."

Trial #2:   Gather the remaining pennies, shake them up, and dump them on your desk. Remove any pennies that have the "tails" side up and record the number of pennies left.

Trial #3 --> Trial #?:   Continue this process until the last penny is removed.

Answer the following questions.

a)   Would the results of this experiment been significantly different if you had removed the "heads" pennies each time?

b)   Would the results have been significantly different if you had alternated heads/tails after each trial?

c)   If you had started with 200 pennies, how would this have effected the results?

d)   How does yesterday's homework question FX-17, about probability, relate to this investigation?

e)   Decide what your dependent and independent variables are, clearly label them, and draw a graph of your data.

f)   Is it possible that some group conducting this experiment might never remove their last penny? Explain.

FX-19   When a radioactive isotope undergoes decay, it does so at a fixed rate. The time it takes for half of it to decay is called the "half-life." For example, if a substance has a half-life of ten years, and you start with 100 grams of it, after ten years you will have only 50 grams. In another ten years, half of that 50 grams will decay, and so on.

years	amount left
0	100
10	50
20	
30	

a)   Make a table of values like the one shown.

b)   Extend the table for values up to 60 years.

c)   Sketch a graph to represent this situation.

d)   How much will be left after 25 years?

e)   When will this isotope disappear completely?

f)   Explain why you think most states will not allow radioactive materials to be dumped inside their borders.

On the next problem, part (f) is best done with a graphing calculator.

FX-20.   A video loses 50% of its value every **year** it is in a video store. The **initial value** of the video was $60.

a)   What is the **multiplier**?

b)   What is the value of the video after one year?

c)   What will the value be after four years?

d)   Write a function $V(t) = ?$ to represent the value in t years.

e)   When does the video have no value?

f)   Sketch a graph of this function. Be sure to scale and label your axes.

FX-21.   Two very unlucky gamblers sat down in a casino with $10,000 each. Both lost every bet they made. The first gambler bet half of his current total at each stage, while the second bet one-fourth of his current total at each stage. They had to quit betting after each of them was down under $1. How many bets did each person get to place?

FX-22.   Find the **annual multiplier** for each of the following.

a)   A yearly increase of 5% due to inflation.

b)   An annual decrease of 4% on the value of a television.

c)   A monthly increase of 3% in the cost of groceries.

d)   A monthly decrease of 2% on the value of a bicycle.

FX-23.   Judy does not believe that x^0 can possibly have any meaning, and if it does, she can't remember it anyway. Kelly volunteers to convince her and starts by showing her the following pattern:

$$\frac{x^7}{x^2} = x^5, \qquad \frac{x^7}{x^3} = x^4, \qquad \frac{x^7}{x^4} = x^3, \qquad \frac{x^7}{x^5} = x^2.$$

a)   Explain how continuing this pattern will help Judy understand x^0.

Kelly helps Judy see the pattern by asking her to simplify each of these fractions:

$$\frac{x^{103}}{x^{98}} \qquad \text{and} \qquad \frac{x^{1052}}{x^{1049}}$$

b)   Simplify these for Kelly.

c)   How would you represent $\frac{x^A}{x^B}$?

Kelly then says that if $\frac{6}{6} = 1$ and $\frac{x^3}{x^3} = 1$, then $\frac{x^B}{x^B} = ...$

d)   Continue with Kelly's explanation. Write it all out so that Judy will understand that $x^0 = 1$.

FX-24.   What are the coordinates of the y-intercepts for each of the graphs below?

a) $y = 2^x$ b) $y = 3^x$

c) $y = 101^x$ d) $y = \left(\frac{3}{5}\right)^x$

e)   State in a sentence any conclusion you can make about the y-intercept for the graph of $y = a^x$?

FX-25.   For $-4 \le x \le 4$ compute both $y = 0^x$ and $y = x^0$

  a)   Record the values on a graph.

  b)   Where do these graphs intersect?

  c)   What is $0^0 = ?$ What does the calculator give for its value? Explain why.

FX-26.   Convert each side of the equation $4^4 = 16^2$ to powers of two in order to show that 4^4 equals 16^2.

FX-27.   Solve each of the following for x. For (c) and (d) use the idea from the previous problem so that you don't have to guess and check.

  a) $2^{x+3} = 2^{2x}$ b) $3^{2x+1} = 3^3$

  c) $9^{40} = 3^x$ d) $8^{70} = 2^x$

FX-28.   If $y - 4 = 4 (x - 3)$,

  a)   what is the slope of this line?

  b)   what is the y-intercept?

  c)   what is the x-intercept?

  d)   Make a quick sketch of this function.

FX-29.   Find the value of x that will make $f(x) = 5$ where $f(x) = \dfrac{6}{x - 1}$.

FX-30.   Solve the following system of equations.

$$5x - 4y = 7$$
$$2y + 6x = 22$$

THE FUNCTION WALK

FX-31.   Your instructor will give you an integer and an index card. Write down the integer on one side of the index card. You will need this card, your calculator, and a pencil for this problem.

Your instructor will give you the rest of the directions.

FX-32.   Make a table like the one below for each of the four functions from the function walk and fill in as much as you can. Give y -values in <u>fraction</u> form and look for a pattern. Divide up the work among group members, but make sure everyone records all the results.

x	2^x	y
4		
3		
2		
1		
0	2^0	
-1	2^{-1}	
-2	2^{-2}	
-3	2^{-3}	
-4		

♀FX-33.   Use the table that you created for $y = 2^x$ to write explanations for the following questions.

a)   Why should $2^0 = 1$?

b)   Why should $2^{-3} = \frac{1}{8}$? What would 3^{-3} equal? 3^{-4} ?

c)   Write an explanation in your Tool Kit of what a negative exponent does.

d)   Should your graphs be **discrete** (points) or **continuous** (connected)?

FX-34.   Draw a graph for each of the four functions from the function walk and give the domain and range. You may want to check your graphs with a graphing calculator.

♀FX-35.   Is your graph for $y = 2^x$ in the problem FX-34 just a set of points or did you draw a smooth continuous curve through all the points?

a)   Make a copy of this graph and draw a smooth curve connecting the points if you have not already done so.

> This function is the simplest example of an **<u>exponential function</u>**. An exponential equation has the general form $y = k(m^x)$, where k is the initial value and m is the multiplier. Be careful. The independent variable x has to be in the exponent. For example, $y = x^2$ is **NOT** an exponential equation, even though it has an exponent.

b)   Add this graph to your Tool Kit, label it EXPONENTIAL FUNCTION, and, on the graph, write the general form for an exponential function.

c)   Have you seen equations of this type before? Where?

FX-36.   Use your graph of $y = 2^x$ and your calculator to answer the following questions.

a)   If $x = \frac{1}{2}$, what value do you get on your calculator (for $y = 2^x$)?

b)   Does this value appear to agree to the corresponding y-value on your graph?

c)   What result do you get for $x = \frac{5}{4}$ on your graph? On your calculator?

FX-37.   Use the graph of $y = 2^x$ to approximate the value of each expression to the nearest tenth.

a) $2^{0.7}$ b) $2^{-0.3}$

c) 2^{-1} d) $2^{1.5}$

FX-38.   Jim does not understand negative exponents. He thinks that 4^{-2} is a negative number. Thu volunteers to help him out. She begins with $\frac{x^3}{x^5}$. Finish writing Thu's explanation, so that Jim will understand the meaning of negative exponents.

FX-39.   Manhattan Island (New York City) was purchased in 1626 for trinkets worth $24. The value of Manhattan in 1994 was about $34,000,000,000. Suppose the $24 had been invested instead at a rate of 6% compounded each year.

a)   What is the **multiplier**?

b)   What is the **initial value**?

c)   What is the **time** (number of years) to find the value of the money in 1994?

d)   Write a function, $V(t) = ?$, to find the **value** 'V' of the investment at any time 't'.

e)   What would the investment be worth in 1994?

f)   How does this value compare to the 1994 value of Manhattan?

FX-40.   Write each expression below with the smallest integer base you can.
(Example: $\frac{1}{16} = 2^{-4}$)

a) $\frac{1}{125}$ b) $(\frac{1}{9})^x$ c) $(\frac{1}{32})^{1-x}$

FX-41.   Solve each equation for x.

    a) $5^x = 5^{-3}$ b) $6^x = 216$

    c) $7^x = \dfrac{1}{49}$ d) $10^x = 0.001$

FX-42.   For each of the following situations, identify the **multiplier**, the **initial value**, and the **time**. Remember that the time must be in the same units as the multiplier. (Example: 3% raise per quarter for two years ---> multiplier = 1.03, time = 8 quarters.)

    a)   A house purchased for $126,000 has lost 4% of its value each year for the past 5 years.

    b)   A 1970 Richie Rich comic book has appreciated at 10% per year, and originally sold for 35¢.

    c)   A Honda Prelude depreciates at 15% per year. Six years ago it was purchased for $11,000.

FX-43.   Solve the following system of equations.

$$y = -x - 2$$
$$5x - 3y = 22$$

FX-44.   Consider the following sequence with initial value 7: $7, 6\frac{1}{3}, 5\frac{2}{3}, 5, \ldots$

    a)   What kind of sequence is it?

    b)   Write an equation that describes the sequence t(n) =

    c)   What is the fifteenth term, t(15)?

    d)   Is -21 a term of the sequence? If so, which one?

FX-45.   State the coordinates of the **x** and **y** intercepts for the graph whose equation is given below.

$$y = -x^2 + 3x - 5$$

FX-46.   An integer between 10 and 20 is selected at random. What is the probability that three is a factor of that integer?

FX-47.   Solve for **x**.

    a) $x^3 = 27$ b) $x^4 = 16$ c) $x^3 = -125$

ROOTS AND FRACTIONAL EXPONENTS

FX-48.   Answer each of the following both numerically and in words

a)   What does the "square root of 25" mean? What does the "square of 25 mean?
What is the difference in the meaning of the two expressions?

b)   What does "8 cubed" mean? What does the "cube root of 8" mean? Explain the
difference in meaning.

c)   What is the meaning of "81 to the fourth power?" What does the "fourth root of
81" mean?

FX-49.   Compute the following without a calculator.

a) $\sqrt[2]{49} =$ b) $\sqrt[4]{16} =$

c) $\sqrt[4]{10000} =$ d) $\sqrt[5]{-32} =$

Write out in words what each of the expressions below says in symbols, then check
them in your head or on your calculator:

e) $13^2 = 169$ and $\sqrt[2]{169} = 13$ f) $7^3 = 343$ and $\sqrt[3]{343} = 7$

FX-50.   Make a table like the one below, fill in any missing information, and complete the table.

POWERS	in other words...	ROOTS	in other words...
$11^2 = 121$			The square root of 121 is 11
$5^3 = 125$	5 cubed is 125	$\sqrt[3]{125} = 5$	The cube root of 125 is 5
$7^4 = 2401$	7 to the 4th is ____		The fourth root of 2401 is 7
			The fifth root of 243 is 3
		$\sqrt[6]{64} =$	
$a^n = b$			

⚲FX-51.   Use a scientific calculator (graphers generally do not have "cube root" buttons) to determine the decimal value of each of the following expressions to the nearest .0001 (ten-thousandth). For the fourth root do the square root twice.

COLUMN A	COLUMN B
RADICAL FORM	EXPONENT FORM
$\sqrt[2]{9}$	$9^{1/2}$
$\sqrt[2]{12}$	$12^{1/2}$
$\sqrt[3]{125}$	$125^{1/3}$
$\sqrt[3]{120}$	$120^{1/3}$
$\sqrt[4]{9}$	$9^{1/4}$

   a)   Compare the values in column A to the values in column B. What conclusions can you draw?

   b)   Using your conclusions, how could you rewrite $\sqrt[x]{b}$?

   c)   Rewrite this relationship as a rule and add it to your Tool Kit.

FX-52.   a)   What is the value of $(\sqrt[3]{50})^3$?   What is the value of $(50^{1/3})^3$?

   b)   Explain why it makes sense for the value of $\sqrt[3]{50}$ to be the same as $50^{1/3}$.

   c)   What exponential expression is the same as $\sqrt[5]{90}$?

FX-53.   Compute the values for the following expressions

   a) $(64^2)^{(1/3)} = ?$ b) $(64^{1/3})^{(2)} = ?$ c) $(64^{(2/3)}) = ?$

   Now compute the values for the following three:

   c) $(625)^{(3/4)} = ?$ d) $(625^{1/4})^{(3)} = ?$ e) $(625^3)^{(1/4)} = ?$

   f)   Write $(32)^{(2/5)}$ in two other ways and calculate the answer in each of the three ways.

   g)   What can you conclude?

FX-54.   Rewrite each of the following. Use fractional exponents instead of radicals and exponents.

   a) $\sqrt[3]{5^2}$ b) $\sqrt[4]{2^5}$ c) $\sqrt[2]{7^3}$

FX-55.   On your calculator evaluate $16^{3/2}$. What two calculations would you have to do to get this result <u>without a calculator</u>?

FX-56.   Solve each equation.

a) $\left(\sqrt[3]{125}\right)^2 = x$ b) $125^{2/3} = x$

c) $\left(\sqrt{x}\right)^3 = 125$ d) $x^{3/2} = 125$

e)   What do you notice about the answers to the above problems? How are the equations related to each other?

FX-57.   Use the ideas from the previous problem to solve the following.

a) $x^{1/4} = 2$ b) $m^{1/3} = 7$ c) $r^{3/2} = 8$

FX-58.   Show two steps to calculate each of the following <u>without</u> a calculator. Be careful, there is an easy way and a hard way for each.

a) $8^{5/3}$ b) $25^{5/2}$ c) $81^{7/4}$

FX-59.   Solve each of the following for x.

a) $2^{1.4} = 2^{2x}$ b) $8^x = 4$ c) $3^{5x} = 9^2$

♀FX-60.   Consider the following pattern: $\dfrac{1}{2^3} = \dfrac{1}{8}$, $\dfrac{1}{2^2} = \dfrac{1}{4}$, $\dfrac{1}{2^1} = \dfrac{1}{2}$, $\dfrac{1}{2^0} = 1$.

a)   What are the values of: $\dfrac{1}{2^{-1}}$, $\dfrac{1}{2^{-2}}$, $\dfrac{1}{2^{-3}}$, $\dfrac{1}{2^{-4}}$?

b)   What is the value of $\dfrac{1}{2^{-n}}$?

c)   Write this as a rule and add it to your Tool Kit (Exponent Laws)

FX-61.   Which of the following do you think are exponential functions? Explain.

a) $f(x) = 3x^2$ b) $y = 2^x + 3^x$ c) $y = 5(4)^x$

d) $g(x) = (2.46)^x$ e) $y = (2^x)(3^x)$ f) $y = (1/2)^x$

FX-62.　Three Algebra 2 students are doing a problem in which they have to find an equation for a pattern. They get three answers: $y = 2^{-x}$, $y = \dfrac{1}{2^x}$, and $y = (0.5)^x$. Explain to these three students what has occurred.

FX-63.　Find the **multiplier** and **time** for each of the following.

　　a)　A yearly increase of 1.23% in population.

　　b)　A monthly decrease of 3% on the value of a video.

　　c)　The annual multiplier if there is a monthly decrease of 3% on the value of a video.

FX-64.　Solve each of the following for x. You can guess and check, or you can be clever and remember such things as $8 = 2^3$.

　　a)　$2^{x+3} = 64$　　　　b)　$8^x = 4^6$　　　　c)　$9^x = \dfrac{1}{27}$

FX-65.　When asked to solve $(x - 3)(x - 2) = 0$, Freddie gives the answer "x = 2." Samara corrects him, saying that the answer should also be x = 3. But Freddie says that when you solve an equation, you only have to find <u>one</u> value of x that works, and since 2 works, he's done. Do you agree with Freddie? Justify your answer.

FX-66.　Solve for z:　$\dfrac{1}{x} + \dfrac{1}{y} = \dfrac{1}{z}$.

FX-67.　On the spinner at right, each "slice" is the same size. What is the probability that when you spin you will get:

　　a)　a one?

　　b)　a two?

　　c)　a three?

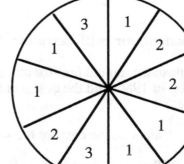

FX-68.　Solve each of the following for x.

　　a)　$(3.25 \times 10^{27})^x = 1$

　　b)　$\left(\sqrt[7]{239^3} \right)^x = 1$

　　c)　$\left(\dfrac{287625}{1191628} \right)^x = 1$

　　d)　$\left[\sqrt[3]{\pi} \left(\dfrac{4}{5} \right)^{-6} \right]^0 = x$

APPLICATIONS OF EXPONENTIAL FUNCTIONS

FX-69.   Below are several situations that can be described by using exponential functions. They represent a small sampling of the situations where quantities grow or decay over some period of time. For each situation (a) through (f):

    i)   Find an appropriate **time** unit (days, weeks, years, etc.).

    ii)   State the **multiplier** that would be used for that situation.

    iii)   Identify the **initial value**.

    iv)   Write an **equation**, in exponential form: $V(t) = k(m^t)$ **, that represents the growth or decay and V(t) represents the value at any given time 't'.

    ** (**m** = multiplier, **t** = time, **k** = initial value)

    a)   A house purchased for $120,000 has an annual appreciation of 6% per year.

    b)   The number of bacteria present in a colony is 180 at 12 noon, and the bacteria grows at a rate of 22% per hour.

    c)   A 100 gram sample of a radioactive isotope decays at a rate of 6% every week.

    d)   The value of a car with an initial purchase price of $12,250 which depreciates by 11% per year.

    e)   For an investment of $1000 the 6% annual interest is compounded monthly.

    f)   For an investment of $2500 the 5.5% annual interest is compounded daily.

FX-70.   Choose one of the exponential equations that you created in the previous problem and create a possible test question that could be solved using the equation.

A Graphing calculator will certainly be useful for part (e) in the next problem.

FX-71.   Suppose the annual fees for attending a campus of the University of California were $1200 in 1986 and the cost increased by 10% each year (round answers to the nearest dollar).

    a)   Calculate the cost for the year 1997.

    b)   What would you expect the cost to be four years from 1997?

    c)   What was the cost in 1980?

    d)   Write an equation to describe this situation.

    e)   Sketch a graph of this function.

    f)   By 1993 the fees totalled $3276. Locate this point in relation to your graphs. Is your model reasonable? Explain.

FX-72.   The California Legislature passed a law that states the cost of fees for the State University system can increase by 5% per year without approval by the public or the legislature (larger increases have to be approved). Annual fees for the university were $1000 in 1993. The Smiths just had a baby girl and wanted to plan for her college education. They figured that, in 18 years, their daughter would be attending one of the state universities. Since 5% of $1000 is $50, they figured the increase in fees in 18 years will be 18 times $50, or $900; therefore, when Kelly starts college the annual fees will be $1900. Explain to the Smiths why they may need to save more than they are anticipating.

FX-73.   According to the U.S. Census Bureau the population here in the United States has been growing at an average rate of approximately 2% per year. The census is taken every 10 years and the population in 1980 was estimated at 226 million people

   a)   How many people should the Census Bureau have expected to count in the 1990 census?

   b)   How many people should the Census Bureau expect to count in the census in the year 2000?

   c)   If the rate of population growth in the U.S. continues to stay at about 2%, in about what year will the population in the United States reach and surpass one billion?

FX-74.   If you can purchase an item that costs $10.00 now and inflation continues at 4% per year (compounded yearly), when will the cost double? Show how you would compute this cost. Remember you can use your calculator to guess and check once you set up an equation.

FX-75.   Solve each of the following for x. Notice the variety in the equations today! Each one requires some different thinking.

   a) $2^{x-1} = 64$ b) $4.7 = x^{1/3}$

   c) $8^{x+3} = 16^x$ d) $9^3 = 27^{2x-1}$

   e) $x^6 = 29$ f) $25^x = 125$

FX-76.   Rewrite $(16)^{3/4}$ in as many different forms as you can.

FX-77.   Write and describe a situation that would fit the following function.

   $g(t) = 0.05(1.16)^t$

FX-78.   For each of the problems below, find the initial value.

   a)   Five years from now, a bond that appreciates at 4% per year will be worth $146.

   b)   Seventeen years from now, Ms. Speedi's car, which is depreciating at 20% per year, will be worth $500.

FX-79.   Factor each expression. An example is given as a reminder.

$$2x^2 - 8x - 42 = 2(x^2 - 4x - 21) = 2(x - 7)(x + 3)$$

a) $x^2 + 8x$ b) $6x^2 + 48x$

c) $2x^2 + 14x - 16$ d) $2x^3 - 128x$

FX-80.   Consider this sequence:

n	0	1	2	3	4
t(n)	10	5	2	1	2

a)   Predict the next three terms.

b)   Graph this sequence. Is it arithmetic, geometric, or something else?

c)   Is your graph discrete or continuous?

FX-81.   In physics, the formula for resistance in a parallel circuit is given by $\frac{1}{r_1} + \frac{1}{r_2} = \frac{1}{R_r}$.
When asked to solve this equation for R_r, Dao wrote: $R_r = r_1 + r_2$. Explain why this is wrong and show how to get a correct solution.

FX-82.   When we say "x = 3" are we talking about a point or a line? Explain.

FX-83.   Graph each of the following. For example in (a) you can find the "edge" of the graph by graphing the line $y = 2x + 3$, then you'll need to decide what area to shade.

a) $y < 2x + 3$ b) $y \geq -3x - 1$

FX-84.   **FAST CARS**

As soon as you drive a new car off the dealer's lot the car is worth less than what you paid for it. This is called **depreciation;** so, you will sell it for less than the price that you paid for it. Some cars depreciate more than others (that is, at different rates), but most cars depreciate. On the other hand, some older cars actually increase in value. This is called **appreciation**. Let's suppose you have a choice between buying a 1993 Mazda Miata for $17,000 which depreciates at 22% a year, or a used 1990 Honda CRX for $12,000 which only depreciates at 18% a year. In how many years will their values be the same? Should you instead buy a 1967 Ford Mustang for $4,000 that is appreciating at 10% per year? Which car will have the greatest value in 4 years? In 5 years?

["Lap" means "part" (because it's about cars).]

Lap 1:   What is the multiplier for the Miata? For the Honda? For the Mustang?

Lap 2:   Make a table like the one below and calculate the values for each car for each year.

year	Mazda Miata	Honda CRX	Mustang
0	$17,000	$12,000	
1	$13,260	9,840	
2			
3			
4			
5			
•			
•			
•			
10			
•			
•			
•			
n			

Lap 3:   Write two functions to represent the depreciation of the Miata and the CRX and draw the graphs on the same set of axes. Are the graphs linear? How are they similar? How are they different?

Lap 4:   When will the values of the two cars be the same?

Lap 5:   Write an equation and graph the results for the Mustang on the same axes with the Miata and the CRX.

For the Checkered Flag: Using the graph, which of the 3 cars -- the Mazda, the Mustang, or the CRX is worth the most after 4 years? after 5 years? after 10 years?

Victory Lap: (Evaluation) After doing this problem what you have learned? Has this problem changed your view of buying cars? Pick one of the three cars and explain why you would buy it.

FX-85.   "Half-life" applies to other situations besides radioactivity. It can apply to practically anything that is depreciating or decaying.

a)   From the table of values in the previous problem, estimate the half-life of the value of the Mazda and the Honda.

b)   According to the mathematical model (not necessarily corresponding to reality), when will each car have <u>no</u> value?

Go back and check on how you solved FX-75 (b) and (e). Be sure your group knows how to do these because you will need this idea today for FX-86 and FX-87.

FX-86.   Find the annual rate of growth on an account that was worth $1000 in 1990 and was worth $1400 in 1993.

FX-87.   Find the monthly rate of decay on a radioactive sample that weighed 100 grams in May and weighs 50 grams in November.

FX-88.   Graph the system of equations below. In the first graph be sure to include a value between 0 and 1 and a value between -1 and 0.

$$y = x^3 \quad \text{and} \quad y = x$$

a)   How many times do these functions cross?

b)   What are the coordinates of their intersections?

c)   Solve the equation $x^3 = x$ for x.

FX-89.   Solve each of the following for x.

a) $25^{x+1} = 125^x$ b) $8^x = 2^5 \cdot 4^4$ c) $27^{(x/2)} = 81$

FX-90.   Factor each expression into three factors

a) $2x^2 + 8x + 8$ b) $6x^2 - 6x - 72$

FX-91.   To what power do you have to raise …

a)   3, to get 27?   b)   2, to get 32?   c)   5, to get 625?

d)   64, to get 8?   e)   81, to get 3?   f)   64, to get 2?

g) (x^3), to get x^1?   h) (x^3), to get x^{12}?   i)   x, to get x^a?

FX-92.   If $f(x) = 3(2)^x$, find

a)   f(-1)   b)   f(0)   c)   f(1)

d)   What value of **x** gives f(x) = 12?

e)   Where does the graph of this equation cross the x-axis? The y-axis?

FX-93.   Sketch a graph for each of the following:

a)   The temperature of a hot cup of coffee left sitting in a room for a long period of time.

b)   The relationship between the amount of money that you have in your savings account and time.

FX-94.   Solve each of the following for **x**. Again, each problem requires different thinking.
a) $81 = 3^{2x}$ b) $x^5 = 243$ c) $(2x)^3 = -216$

FX-95.   Rewrite each of the following in y-form so that they are in the correct format for use on a graphing calculator.

a)   2x - 3y = 7

b)   2(x + y) = x - 4

FX-96.   Solve each of the following for **x**.

a) $\dfrac{x + 3}{x} = \dfrac{4}{5}$ b) $\dfrac{5}{x} = \dfrac{x}{10}$

FX-97.   Solve the following systems of equations.

a)   2x + y = -7y   b)   3s = -5t
      y = x + 10   6s - 7t = 17

FX-98.   Sketch the graphs of three different parabolas with x intercepts at (4, 0) and (8, 0). What can you tell about the vertex of each parabola?

FX-99.   Solve the following system of equations for D, E, and F. Think of it as a puzzle.

$$F = -5ED^2$$
$$D = 3$$
$$6E = 2F - 32$$

FX-100. The area of square A is 121 square units, the perimeter of square B is 80 units. Find the area of square C.

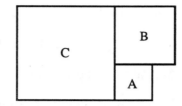

TWO FUNCTION INVESTIGATIONS

FX-101. Investigate the family of functions $y = a^x$ for different values of "a." Be sure that you tried enough different values of "a."

FX-102. Two weeks ago a sample of bacteria weighed 4.2 grams. Last week it weighed 4.326 grams. Figure out the multiplier.

   a)   What is the rate of growth?

   b)   What is the weight **now**?

FX-103. Last month the Portia's car was worth $28,000. Next month it will be worth $25,270.

   a)   What is the rate of depreciation?

   b)   What is the car worth now?

FX-104. Investigate the family of functions $y = ax^2$. Be sure to use both negative and positive numbers for "a." You'll need the results of this investigation for the beginning of the next chapter.

FX-105.   **CHAPTER 3 SUMMARY ASSIGNMENT**

   In this chapter you have worked with exponential functions in many contexts.

   a)   Make up a real situation that can be represented by a exponential function.

   b)   Give the equation that describes the situation and tell how you arrived at the equation.

   c)   Draw the graph of the equation. Be sure to include all important information about the graph. How did you decide if the points on the graph should be connected or not?

   d)   Make up a question that can be solved by using your equation or your graph. Clearly show how to answer the question.

💡FX-106.   Now would be a good time to review your Tool Kit and make sure you have included any new ideas about exponents that you learned in this chapter. When you have included new rules (such as $b^{1/x} = \sqrt[x]{b}$) which were based on patterns, write an explanation of why the rule should be true and discuss whether it is true for all possible choices of the variables or are there some limitations?

FX-107.   Many Algebra 1 students think that $2^{-2} = -4$. However you know that $2^{-2} = \dfrac{1}{2^2} = \dfrac{1}{4}$.

Explain, so that an Algebra 1 student can understand, why $2^{-2} = \dfrac{1}{4}$

FX-108.   Jonnique is writing a puzzle problem. She wants the values for **x** and **y** in the second equation to be the same as in the first. She originally wanted the values to be whole numbers so they could be guessed and checked, but the whole numbers she has tried don't work. Show a method, other than guess -and-check, for figuring out what numbers will work.

Jonnique's equations: $2^x \cdot 2^y = 64$ and $2^{3x} = 16^y$

FX-109.   A sequence is given by $t(n) = 2(3)^n$.

a)   What are t(0), t(1), t(2), t(3)?

b)   Graph this sequence. What are the domain and range?

c)   On the same set of axes, graph the function $f(x) = 2(3)^x$.

d)   How are these graphs similar? How are they different?

FX-110.   Solve the following for **x**. You'll need to guess and check for part (b).

a) $1^x = 5$ b) $2^x = 9$

FX-111.   Solve each of the following for **x** and **y** where **x** and **y** are whole numbers. The fact that the prime factorizations of both 72 and 24 are all 2's and 3's should help.

a) $2^x 3^y = 72$ b) $2^{x+y} 3^{x-y} = 24$

FX-112.   Write each of the expressions below in an equivalent form without negative or fractional exponents and then give the value. You should be able to do all of these problems without your calculator!

a) 5^{-2} b) 4^{-3}

c) $9^{1/2}$ d) $64^{2/3}$

FX-113.   You have $5000 in an account that pays 12% annual interest. Compute the amount in the account at the end of one year if

a)   the interest is paid annually (once)

b)   the interest is paid quarterly

c)   the interest is paid monthly

FX-114.   Find an equation of decay of a radioactive sample in the form $f(t) = km^t$ if $f(0) = 100$ and $f(5) = 50$

FX-115.   Rewrite each equation in y-form and then find the point(s) of intersection.

$$2x - 5y = 10 \text{ and } 4(x - y) + 12 = 2x - 4$$

FX-116.   Draw a quick sketch of the relationship between your test scores and the amount of homework that you do.

FX-117.   Find the coordinates of the x- and y-intercepts and the equation for the line of symmetry for the following function.

$$y = x^2 + 14x + 13$$

EXPONENTS TOOL KIT

Dividing Like Bases

Exponential Equations

Fractional Exponents

Multiplying Like Bases

Negative Exponents
(Reciprocals)

Powers of Powers

Zero Power

Chapter 4

Parabolas and Other Parent Graphs
(The Gateway Arch)

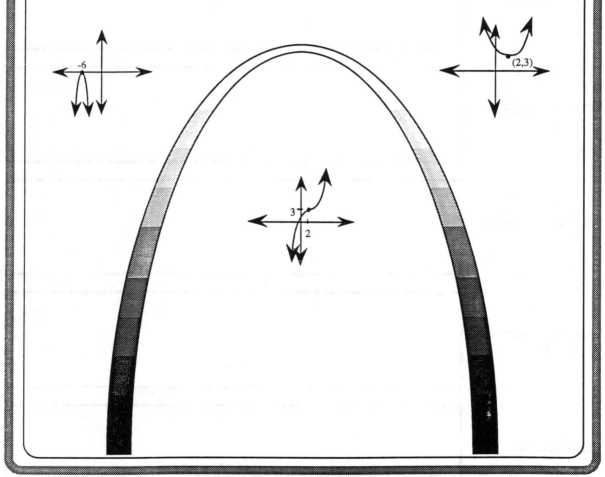

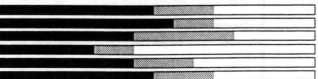

CHAPTER 4
THE GATEWAY ARCH:
PARABOLAS AND OTHER PARENT GRAPHS

In Chapter 4, you will have the opportunity to:

- become very familiar with an extended set of parent equations and their graphs. These will include the parabola, absolute value, and circle, a cubic and a hyperbola.

- relate the numbers in an equation to the location and stretch of each parent graph.

- develop your ability to make generalizations about functions as you move from specific examples to more abstract general equations.

- develop your ability to ask questions to generate more information and to justify and explain your reasoning.

You will use your graphing calculator daily as a tool to help you experiment and discover. Just as in science, you will write up your observations and conclusions in lab reports. The chapter ends with several large, challenging investigations.

THE GATEWAY ARCH AND PARABOLA LAB #1

PG-1.   Would you ever go bungee jumping? Beginners jump off bridges over water, or off cranes over a big air cushion. But true daredevils (some call them idiots) dream of jumping off the Gateway Arch in St. Louis, Missouri, which has no soft landing area in case the bungee cord breaks. No one has ever done this, and we are not recommending it, but it makes a challenging problem.

The most daring jumper might jump off the very top of the Arch, but this is 185 meters high. Jumpers that are chicken (or sensible) might prefer to start at a lower point on the Arch. We want a rule that will tell us how high the Arch is at any point, so they'll know how long a bungee cord to use.

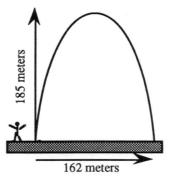

a)   What kind of curve could be used to approximate the Arch? That is, what kind of equation has a graph with a shape like this?
      [**a parabola or quadratic**]

b)   The Arch is approximately the shape of an upside-down parabola. What is the simplest equation for a parabola that you know of?   [$y = x^2$]

c)   Think back to when you investigated $y = ax^2$. How can we change the equation of the simple parabola from part (b) to get an equation to represent the Arch? This is a tough problem. Discuss your ideas with your group and be ready to share them with the class in a few minutes.   [**see teacher note below**]

PG-2.   **PARABOLA LAB #1**

1. Graph the equation $y = (x - 2)(x - 2)$ using your graphing calculator. What kind of curve is it? Make a sketch of the graph. Be sure to label any important points.

2. Use your calculator to find two other (different) parabolas that "sit" on the x-axis at $x = 2$. They should open up, and should touch the x-axis at only one point (that is, the point $(2, 0)$). Keep track of your observations as you try different parabolas, even the ones that don't work.

    Make sketches of the two graphs and label them with their equations. Compare these new parabolas to the one in part (1).

3. Use your calculator to find two different parabolas that hang **down** from the x-axis at $x = 2$. They should open down, and also should touch the x-axis at only one point.

    Sketch the two graphs and label them with their equations.

4. Sketch a parabola that hangs down from the x-axis at $x = -4$. Record the equation you used. Then find another equation whose graph hangs down from the same point and graph it.

    Sketch these two graphs and label them with their equations.

5. Make predictions as to how many places the graph of each equation below will touch the x-axis. You may want to first factor some of the equations into a more useful form.

    a) $y = (x - 2)(x - 3)$ b) $y = (x + 1)^2$ c) $y = x^2 + 6x + 9$

    d) $y = x^2 + 7x + 10$ e) $y = x^2 + 6x + 8$ f) $y = -x^2 - 4x - 4$

6. Check your predictions with your calculator.

7. Write a clear explanation describing how you can tell whether the equation of a parabola will touch the x-axis at only one point.

8. In the equations below, fill in the blanks with the numbers that would make the resulting equation the equation of a parabola that touches the x-axis at only one point. Experiment with your calculator to find the right value.

    a) $y = x^2 + 4x +$ _____   b) $y = x^2 + 6x +$ _____

    c) $y = x^2 + 8x +$ _____   d) $y = x^2 +$ _____

    e) $y = x^2 + 5x +$ _____   f) $y = x^2 - 2x +$ _____

    g) $y = -x^2 - 6x -$ _____

9. **Lab Report:** When you and your partner (group) are finished, write up a lab report. Your instructor will give you a suggested format for your lab report.

                                (Based on an activity by Anita Wah and Henri Picciotto)

PG-3.   Consider the sequence with the initial value 256, followed by 64, 16, . . .

   a)   Write the next three terms of this sequence. Then find a rule.

   b)   If you were to keep writing out more and more of the sequence, what would
         happen to the terms?

   c)   Sketch a graph of the sequence. What happens to the points as you go further to
         the right?

PG-4.   Write the equation for each graph:

   a)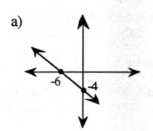

   b)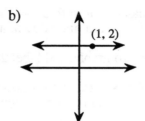

   c)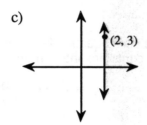

   d)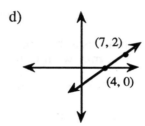

PG-5.   Imagine spinning the rectangle in the diagram about
         the y-axis.

   a)   Draw the shape you'd get.

   b)   Find the volume.

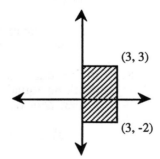

PG-6.   People who live in isolated or rural areas often have their
          own tanks for gas to run appliances like stoves, washers,
          and water heaters. Some of these tanks are made in the
          shape of a cylinder with two hemispheres on the ends
          (a hemisphere is half a sphere, or half a ball).

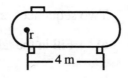

The Insane Propane Gas Tank Company (motto: *"We're Crazy About Gas"*) wants to
make tanks with this shape, but make different models of different sizes. All tanks will
have the cylinder part be 4M long. But the radius r will vary among different models.
Remember the volume of a sphere $V = (4/3)\pi r^3$ and the volume of a cylinder $V = \pi r^2 h$.

a)   One of their tanks has a radius of 1m. What is its volume?

b)   When the radius doubles (to 2m), will the volume double? Explain. Then figure
       out the volume of the larger, 2m tank.

c)   Write an equation that will let Insane Propane Gas Tank Company determine the
       volume of a tank with any size radius.

PG-7.   Remember function machines? Each of the following pictures shows how the same
          machine changes the given x-value into a corresponding f(x)-value. Find the rule for
          this machine:

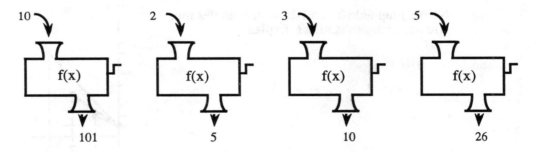

PG-8.   Factor each of the following:

a) $4x^2 - 9y^2$ b) $8x^3 - 2x^7$

c) $x^4 - 81y^4$ d) $8x^3 + 2x^7$

PG-9.   Two sequences are defined as follows: $t(n) = 3n$ and $s(n) = 2^n$.

a)   Fill in the table for each sequence:

n	t(n)	n	s(n)
0		0	
1		1	
2		2	
3		3	

b)   Roger graphed the sequence $t(n)$ (see diagram). He noticed that he could form a sequence of right triangles using $(0, 0)$ that increase in size. Roger thinks that the triangles are similar. Is he right ? Justify your answer.

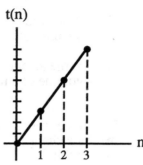

c)   Nicola graphed $s(n)$ and drew in triangles too. Are her triangles similar? Explain.

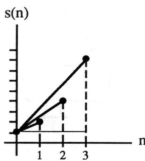

PG-10.   Find the point where $y = 3x - 1$ intersects $2y + 5x = 53$.

PG-11.   Right now, it probably costs an average of $150 each month for your food..

a)   Five years from now how much will you be spending on food each month if you're eating about the same amount and inflation runs at about 4% per year?

b)   Write an equation that represents your monthly food bill x years from now if both the rate of inflation and your eating habits stay the same.

PG-12.   **PARABOLA LAB #2**

1.   Graph the parabola $y = x^2$. Make an accurate sketch of the graph. Be sure to label any important points on your graph.

2.   Find a way to change the equation to make the same parabola _open downward_. The new parabola should be congruent to $y = x^2$, with the same vertex, except it should open downward.

      Record the equations you tried, along with their results (even if they were wrong--they may come in handy below).

3.   Find a way to change the equation to make the $y = x^2$ parabola _stretch vertically_ (it will appear narrower). The new parabola should have the same vertex and orientation (i.e., open up) as $y = x^2$.

      Record the equations you tried, along with their results.

4.   Find a way to change the equation to make the $y = x^2$ parabola _compress vertically_ (it will appear wider).

      Record the equations you tried, along with their results.

5.   Find a way to change the equation to make the $y = x^2$ parabola _move 5 units down_. That is, your new parabola should look exactly like $y = x^2$, but the vertex should be at (0, -5).

      Record the equations you tried, along with their results.

6.   Find a way to change the equation to make the $y = x^2$ parabola _move 3 units to the right_. That is, your new parabola should look exactly like $y = x^2$, except that the vertex should be at the point (3, 0).

      Record the equations you tried, along with their results.

7.   Finally, find a way to change the equation to make the $y = x^2$ parabola _vertically compressed (wider), open down, move 6 units up and move 2 units to the left_. Where is the vertex of your new parabola?

      Record the equations you tried, along with their results.

8.   For each equation below, predict the <u>vertex</u>, the <u>orientation</u> (open up or down?), and tell whether it is a <u>vertical stretch</u> or <u>vertical compression</u> of $y = x^2$. Sketch a quick graph based on your predictions.

      a) $y = (x + 9)^2$ b) $y = x^2 + 1$ c) $y = 3x^2$

      d) $y = \frac{1}{3}(x + 1)^2$ e) $y = \frac{5}{2}(x - 7)^2 + 6$ f) $y + 1 = 2(x - 2)^2$

      g) $y = 2(x^2 + 6x + 9)$ h) $4y = -4x^2$ i) $y = 4x - 4$

9.   Check your predictions above on your graphing calculator. If you made any mistakes, correct them and briefly describe why you made the mistake (what incorrect idea you had). Then make a neat and accurate graph for each function.

10.   **Lab Report:** When you and your partner (group) are finished, write up a lab report. Use the lab report format suggested by your teacher.

♀PG-13.   Take out your Tool Kit. Write some notes for yourself on how to move, flip, and stretch/compress a parabola. Include a **general equation** for all parabolas.

PG-14.   Solve for z:

a) $5^{(z+1)/3} = 25^{1/z}$

b) $4^{2z/3} = 8^{z+2}$

PG-15.   Leti just got her driver's license! Her friends soon nicknamed her "Leadfoot" because she's always doing 70 mph on the freeway.

a)   At this speed, how long will it take her to travel 50 miles?

b)   How long would it take her if she slowed down to 55 mph.?

c)   Speeding tickets carry fines of about $200 and usually raise the cost of insurance. Is it worth it?

PG-16.   Daniela, Kieu, and Duyen decide to go the movies one hot summer afternoon. The theater is having a summer special: Three Go Free (if they each buy a large popcorn and a large soft drink). They take the deal and end up spending $19.50. The next week, they go back again, only this time, they each pay $2.50 to get in, they each get a large soft drink, but they share one large bucket of popcorn. This return trip also costs them a total of $19.50.

a)   Find the price of a large soft drink and the price of a large bucket of popcorn.

b)   Did you write two equations or did you use another method? If you used another method, write two equations now, and solve them. If you already used a system of equations, skip this part.

PG-17.   Multiply each of the following:

a) $2x^3(3x + 4x^2y)$

b) $(x^3y^2)^4(x^2y)$

c.)   There are several subproblems to do before "multiplying." Factor, factor, factor.

$$\frac{x^2 - 9}{2x} \cdot \frac{x^2 + x}{x^2 - 2x - 3}$$

PG-18.   Solve for x: $ax + by^3 = c + 7$

PG-19.   Given $g(x) = 2(x + 3)^2$:

a)   find: $g(-5)$

b)   find: $g(a + 1)$

c)   If $g(x) = 32$, figure out what number x can be.

d)   If $g(x) = 0$, figure out what number x can be.

PG-20.   When a pregnant woman goes into labor, her contractions start out short and mild. These contractions grow progressively in intensity and duration. When they start, the contractions are far apart (30 minutes to an hour) and at the time of delivery, they are happening one right after the other. Sketch a graph showing the relationship between the length of time between contractions and the intensity of the contractions.

PG-21.   Solve for the indicated value.

a)   x = _____   b)   BC = _____

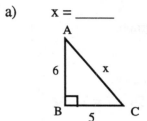   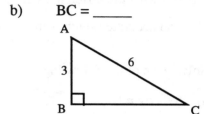

PG-22.   You are making a dart board. It is a square 30 cm on a side. Place a 5 cm circle as a target somewhere within the square to maximize the probability that it will be hit when a dart hits the board at **random**. Where do you plan to place the circle? Explain your thinking about the placement of the circle.

AVERAGING THE INTERCEPTS

PG-23.   Consider the function $y = x^2 - 8x + 7$:

a)   What are the coordinates of the x-intercepts?

b)   Find the average value of the x-intercepts.

c)   What is the y-value that corresponds to this average x-value?

d)   Make a sketch of this function and label the intercepts, the vertex, and draw in a line of symmetry.

e)   Rewrite the equation in the form you found in the parabola lab. What do you notice?

PG-24.   For each of the following two functions, answer each of the questions (a) through (f):

$$f(x) = x^2 + 3x - 10 \qquad\qquad g(x) = x^2 - 4x - 2$$

a)   What are the coordinates of the x-intercepts?

b)   Find the average value of the x-intercepts.

c)   How is the answer to part (b) related to the number in front of **x**?

d)   What is the y-value that corresponds to each average x-value?

e)   Make a sketch of each function and label the intercepts and the vertex, then draw in a line of symmetry.

f)   Rewrite each equation in the form you found in the parabola lab.

g)   Discuss with your group any relationships you see between the average of the intercepts, the equations, and your graphs.

PG-25.   Based on your work in PG-23 and 24 describe a possible method for finding the vertex
         without drawing the graph.

         a)   Give another example, and show how to use your method.

         b)   Use the description you wrote to find the vertex for $y = x^2 + bx + c$.

PG-26.   Explain what the differences are between:

         (1)   an accurate sketch   and   (2)   a careful graph.

PG-27.   Solve each of the following:

         a) $y^2 - 6y = 0$ b) $n^2 + 5n + 7 = 7$

         c) $2t^2 - 14t + 3 = 3$ d) $\frac{1}{3}x^2 + 3x - 4 = -4$

         e)   What do all of the above equations have in common?

PG-28.   Find the vertex of each of the following parabolas by averaging the x-intercepts.

         a) $y = (x - 3)(x - 11)$ b) $y = (x + 2)(x - 6)$

         c) $y = x^2 - 10x + 16$ d) $y = (x - 2)^2 - 1$

♀PG-29.   In the previous problem, part (d) is written in a useful format.

         a)   What are the coordinates of the vertex for part (d)?

         b)   How do these coordinates relate to the equation?

         ┌───┐
         │ The **vertex** of a parabola locates its position on the axes. The **vertex** serves as a │
         │ **Locator Point** for a parabola. The other shapes we will be investigating in this │
         │ course also have locator points. These points have different names but the same │
         │ purpose for each different type of graph.   │
         └───┘

         c)   Add **Locator Point** to your Tool Kit. Use a parabola and its vertex as one
              example. You may wish to leave room for another example if you find one later.

PG-30.   Rewrite each of the following equations so that you could enter them into the graphing
         calculator.

         a) $x - 3(y + 2) = 6$ b) $\frac{6x - 1}{y} - 3 = 2$

         c) $\sqrt{y - 4} = x + 1$ d) $\sqrt{y + 4} = x + 2$

         e)   Find the x and y intercepts for each of the graphs created by the calculator forms
              of the equations in parts a) through d).

PG-31.   Can the quadratic formula be used to solve $4x^3 + 23x^2 - 2x = 0$? Show how or explain why you can't.

PG-32.   Scientists can estimate the increase in carbon dioxide in the atmosphere by measuring increases in carbon emissions. In 1991 the annual carbon emission was about seven gigatons (a gigaton is a billion metric tons). Over the last several years annual carbon emission has been increasing by one percent .

a)   At this rate how much carbon will be emitted in 2000?

b)   Write a function $C(x)$ to represent the amount of carbon emitted in the year 2-thousand-X.

PG-33.   Harvey's Expresso Express, a drive through coffee stop, is famous for its great house coffee, a blend of Colombian and Mocha Java. Their arch rival, Jojo's Java, sent a spy to steal their ratio for blending beans. The spy returned with a torn part of an old receipt that showed only the total number of pounds and the total cost, 18 pounds at $92.07. At first Jojo was angry, but then he realized that he knew the price per pound of each kind of coffee ($4.89 for Colombian and $5.43 for Mocha Java), and that was enough. Show how he could use equations to figure out how many pounds of each type of beans Harvey's used.

PG-34.   What is a *line of symmetry* ?

a)   Draw something that has a line of symmetry.

b)   Draw something that has <u>two</u> lines of symmetry.

c)   Can you find a basic geometric shape that has an <u>infinite number</u> of lines of symmetry?

PG-35.   Lilia wants to have a circular pool put in her back yard. She wants the area around the pool to be cement. If her yard is a 50 ft. by 30 ft. rectangle:

a)   what is the largest radius pool that can fit in her yard?

If the cement is to be 8 inches thick, and costs $1.00 per cubic foot:

b)   what is the cost of putting in the cement?
       (Reminder: Volume = (Base Area) • Depth)

PARABOLAS AS MATHEMATICAL MODELS

PG-36.   We're surrounded by parabolas (help!). Parabolas are good models for all sorts of
things in the world. Indeed, many animals and insects jump in parabolic paths.

This diagram shows a jackrabbit jumping over a
three foot high fence. In order to just clear the
fence, the rabbit must start its jump at a point 4 feet
from the fence.

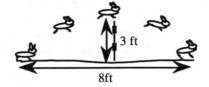

a)   Decide, as a group, where to put the x and y axis on this diagram.

b)   Find the equation of a parabola that models the jackrabbit's jump.

c)   What do the dependent and independent variables represent in this situation?

d)   What are the range and domain of your equation? What parts of the domain and
range are appropriate for the actual situation?

PG-37.   A fireboat in the harbor is assisting in
putting out a fire in a warehouse
along the pier. Use the same process
as in the previous problem to find the
equation of the parabola that models
the path of the water from the fireboat
to the fire, if the distance from the
barrel of the water cannon to the roof
of the warehouse is 180 feet and the
water shoots up 30 feet above the
barrel of the water cannon.

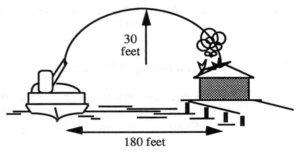

PG-38.   Draw accurate graphs of $y = 3x - 5$, $y = 3x^2 - 5$ and $y = \frac{1}{3}x^2 - 5$ on the same set of axes.

a)   In the equation $y = 3x - 5$, what does the 3 tell you about the graph?

b)   Is the 3 in $y = 3x^2 - 5$ also the slope? Explain.

PG-39.   Do the sides of a parabola ever curve back in like the figure at
right? Give a reason for your answer.

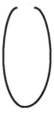

PG-40.   Do the sides of the parabola approach straight vertical lines as shown in the picture at right (in other words, do parabolas have asymptotes)? Give a reason for your answer.

PG-41.   Find the x and y intercepts of:

a) $y = 2x^2 + 3x - 5$ b) $y = \sqrt{2x - 4}$

PG-42.   Sketch graphs. Remember a sketch includes the labelled locator point.

a) $y = 3x^2 + 5$ b) $f(x) = -(x - 3)^2 - 7$

PG-43.   If $g(x) = x^2 - 5$, find:

a) $g(0.5)$ b) $g(h + 1)$

PG-44.   If $g(x) = x^2 - 5$, find the value(s) of x so that:

a) $g(x) = 20$ b) $g(x) = 6$

PG-45.   While watering your outdoor plants, you notice that the water coming out of your garden hose follows a parabolic path. Write the equation of a parabola that describes the path of the water from the hose to the plant.

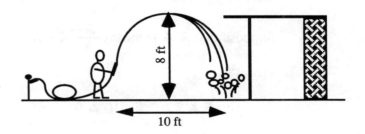

PG-46.   Andrew likes to make up weird sequences and name them after himself. He made up these two: $A(n) = 2(n - 1) + 3$ and $a(n) = 2(n - 1)^2 + 3$.

a)   Are his sequences arithmetic, geometric, or something else?

b)   What would their graphs look like?

PG-47.   Write an equation of a parabola that fits this description:

domain: $-\infty < x < \infty$, and range: $y \geq 2$.

PG-48.   Solve each equation for **x** (that is, put in "x =" form):

a) $y = 2(x - 17)^2$

b) $y + 7 = \sqrt[3]{x + 5}$

PG-49 $\triangle ABC \sim \triangle AED$ in the diagram shown.

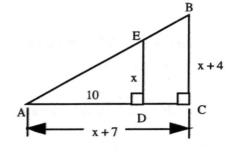

a)   Solve for x.

b)   Find the perimeter of $\triangle ADE$

PG-50.   You are standing on the corner, waiting to cross the street, when you hear a booming car stereo approaching.

a)   Sketch a graph that shows the relationship between how far away from you the car is and the loudness of the music.

b)   Which is the dependent variable and which is the independent variable?

PG-51.   Writers of standardized tests sometimes write problems with weird symbols in them in order to try to confuse you. Show those people that you can't be fooled by solving the following problem:

Suppose that A ♣ B means $\dfrac{A^2 - B^2}{A - B}$.

a)   Find:

     i) 5 ♣ 3   ii) 10 ♣ 9   iii) 28 ♣ 2   iv) 17 ♣ 17   v) x ♣ y

b)   Write a conjecture about A ♣ B. Then test your conjecture with 5 ♣ 5. Does it work? Revise your conjecture if necessary.

c)   How could you show algebraically that your conjecture works for almost any value of A and B?

MOVING OTHER FUNCTIONS

PG-52.   Each person will need two full sheets of graph paper (you will use the front side only). Divide each sheet into six equal sections like this:

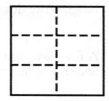

Working as a group, sketch a graph of each of these equations and identify each graph with its equation. Label the coordinates of intercepts or any other important points that are relatively easy to find. You may split up the work in any way you think is fair, but each person should sketch his or her own record of each graph with its equation.

a) $y = x^3$

b) $y = 2^x$

c) $y = \dfrac{1}{x}$

d) $y = \dfrac{1}{x} - 4$

e) $y = (x - 1)^3$

f) $y = 2^x - 4$

*g) $y = 2^{(x-4)} - 3$

*h) $y = \dfrac{1}{x + 3} + 1$

*i) $y = \dfrac{2}{x - 3}$

j) $y = (x - 2)^3 + 1$

k) $y = 2^{(x+3)}$

l) $y = \dfrac{1}{2}(x + 2)^3$

* Be sure to use parentheses when entering these equations in the calculators.

PG-53.   Separate (cut) your graphs along the fold lines and then discuss possible sorting methods with the others in your group. Once your group has decided on a sorting method, sort the graphs into appropriate stacks.

a)   How did you decide which graph went in which stack? What are the similarities and differences among the graphs in each stack?

b)   What are the similarities and differences among the <u>equations</u> of the graphs in each stack?

c)   Which graph, in each pile, has the simplest equation?

d)   Each person in your group should carefully graph the simplest equation from each stack to keep for themselves (keep these graphs with your answers to this problem). You will each need these graphs later in this chapter.

PG-54.   Look at your graph of $y = x^3$. This curve, called a "**cubic**," has an important point at the origin. As you will find out in Calculus, points like these are called "inflection points." If you are curious, ask your instructor for a definition of an inflection point. Find all the other graphs that have inflection points. Write down their equations and the coordinates of their inflection points.

PG-55.   Find all of the graphs that have asymptotes. Write down their equations, along with the equations of the asymptotes. If a graph has two asymptotes that intersect each other, write down the point where they intersect.

PG-56.   A few days ago you explored how to change the equation of $y = x^2$ to shift the graph up and down, left and right. Explain the relationship between that lab and today's graphs.

PG-57.   Sketch a graph of each of these equations (include the y-intercept and "locator point" for each; approximate the x-intercepts where appropriate):

a) $y = x^3 + 5$ b) $f(x) = (x - 10)^3$ c) $g(x) = \dfrac{1}{x} + 7$

d) $y = \dfrac{1}{x + 8}$ e) $y = 2^x + 7$ f) $h(x) = (x - 2)^3 - 6$

PG-58.   The graph of $y = x^2$ is shown in dashed lines. Estimate the equations of the two other parabolas.

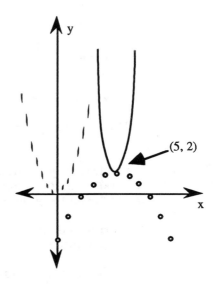

(5, 2)

PG-59.   Use the locator points to write a possible equation for each graph:

a)   b)   c)

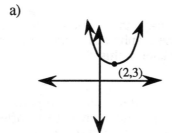

(2,3)

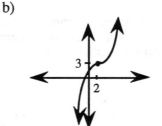

3

2

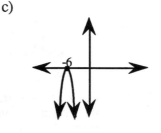

-6

PG-60.   Find the domain and range for each of the graphs in the previous problem.

PG-61. Find the x and y intercepts and the vertex (locator point) of $y = x^2 + 2x - 80$. Then sketch a graph.

PG-62. If $h(x) = (x + 2)^{-1}$, find:

a) h(3) b) h(-3) c) h(a - 2)

PG-63. Multiply each of the following to remove parentheses:

a) (x - 1)(x + 1) b) 2x(x + 1)(x + 1) c) (x - 1)(x + 1)(x - 2)

PG-64. Find the x- and y-intercepts of: $y = (x - 1)(x + 1)(x - 2)$.

PG-65. The two rectangles shown to the right are similar. Find the width of each.

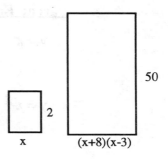

PG-66. Kristin's grandparents started a savings account for her when she was born. They invested $100 in an account that pays 8% interest compounded annually.

a) Write an equation to express the amount of money in the account on Kristin's x^{th} birthday.

b) How much is in the account on her 16th birthday?

c) What is the domain and range for the equation that you wrote in (a)?

PARENT GRAPHS

♀PG-67.   What is the simplest equation of a parabola you can think of?

Most people would probably say that $y = x^2$ is the simplest parabola. We use the term **PARENT GRAPH** to describe the simplest version of a whole family of equations. So $y = x^2$ is the **PARENT** for all parabolas. We can get the equation of any other parabola just by moving, flipping, or stretching $y = x^2$.

a)   What do you think is the parent for all lines? All exponentials?

Some families of functions are so important that they will come up over and over again throughout this course. So it will come in handy to have a **PARENT GRAPH TOOL KIT** -- a special Tool Kit in which to keep a catalog of parents of important equations and shapes (there is a parent graph resource page at the end of this chapter which you can either photocopy or use as an outline). The parents we have seen so far are as follows:

Parent Equation	**Family**
$y = x$	**Lines**
$y = x^2$	**Parabolas**
$y = x^3$	**Cubics** (This is just one type of cubic; there are other cubic equations that don't have $y = x^3$ as a parent -- you will see these in Chapter 7)
$y = 2^x$	**Exponential** (This one needs some discussion because $y = 2^x$ is not really the parent for an exponential equation in base 3. You may decide $y = b^x$ makes a better parent.)
$y = \dfrac{1}{x}$	**Hyperbolas** (there are other parent hyperbolas, also)

b)   Put the information about each of these functions on a **PARENT GRAPH TOOL KIT** page. For each one, include its name, a careful graph that is clearly labeled, a domain and range, and all the other things involved in investigating a function.

PG-68.   Picky Patrick doesn't think that $y = 2^x$ should be the parent of all exponentials. "The parent is supposed to be the *simplest* equation you can get," Patrick says. " I think either $y = 1^x$ or $y = 0^x$ is simpler than $y = 2^x$, so one of those two should be the parent of all exponentials. What do you think?

PG-69.   For each function below:

i)   state the equation of the parent.

ii) create a reasonable sketch of the graph, including intercepts (you should try this without a graphing calculator, but feel free to use one to check your ideas).

iii) give the domain and range.

iv) find the coordinates of the vertex or other locator point, if any.

v)   indicate the location of any asymptotes (and where they intersect each other).

a) $y + 1 = x^2$ b) $y + 4 = x^3$ c) $y + 6 = (x + 2)^3$

d) $y + 2 = 2^x$ e) $y = \dfrac{1}{x - 3}$ f) $y = \dfrac{1}{x} + 2$

g) $y - 3 = (x - 4)^3$ h) $y + 2 = (x - 3)^2 + 1$ i) $y - 5 = 2^{(x + 4)}$

**Note: these may be easier to figure out if you put them all in y-form.

PG-70.   Write a possible equation of each of these graphs. Assume that one mark on each axis is one unit. In class, tomorrow, check your equations on the graphing calculators or compare with members of your group.

a)   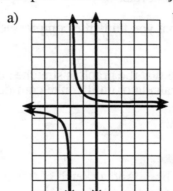   b)   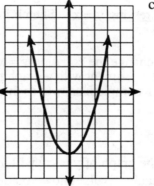   c)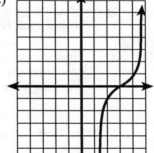

d)      e)      f)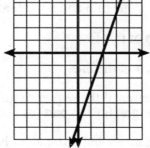

>>>PROBLEM CONTINUES ON THE NEXT PAGE>>>

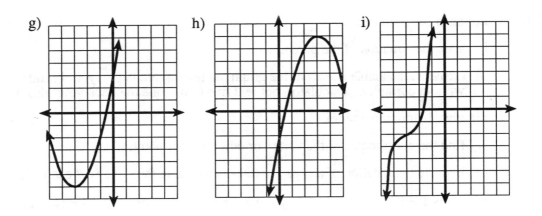

PG-71.   Write the equations of three different parabolas with vertex at (4, 0).

PG-72.   Now that you have had some experience with graphs that are generated from $y = x^2$, we will look at graphs that come from $y = x^3$.

a)   Draw an accurate graph of $y = x^3$. (x-values from -2 to 2 will be sufficient!)

b)   Now imagine <u>stretching</u> this graph just like you stretched a parabola. Sketch in what you think the stretched graph would look like.

c)   Imagine <u>compressing</u> $y = x^3$. Sketch in what you think the compressed graph would look like.

d)   How can you change the equation $y = x^3$ so that the graph is stretched or compressed? Experiment with a graphing calculator until you're confident that you can write a complete answer to this.

PG-73.   Now for $y = -x^3$:

a)   Predict what the graph of $y = -x^3$ will look like. Add a sketch of this function to your $y = x^3$ graph. Explain how you decided what this function will look like.

b)   Graph $y = -x^3$ on your calculator. How does this graph compare with your sketch?

PG-74.   Consider a line with a slope of 3 and a y-intercept at (0, 2).

a)   Sketch the graph of this line.

b)   Write the equation of the line.

c)   Find the first 4 terms of the sequence $t(n) = 3n - 1$. Plot the terms on a new set of axes next to your graph from part (a) above.

d)   Explain the similarities and differences between the graphs and equations in part (a) - (c). Are both continuous?

PG-75.   Caren wrote the following solution, and her teacher marked it wrong but didn't tell her how to answer the problem correctly. Write an explanation of what is wrong and show Caren how to solve the problem .

     X $(x + y)^2 = x^2 + y^2$

PG-76.   The gross national product (GNP) was 1.665×10^{12} dollars in 1960. If the GNP increased at the rate of 3.17% per year until 1989:

    a)   What was the GNP in 1989?

    b)   Write an equation to represent the GNP **t** years after 1960, assuming that the rate of growth remains constant.

    c)   Do you think the rate of growth remains constant? Explain.

PG-77.   Is $y = x^2$ the parent of $y = 9 + 6x - x^2$? Explain how you decided.

PG-78.   Find the coordinates of the intercepts for each of the following functions:

    a) $g(x) = (x + 3)^3$ b) $y - 1 = 3^x$

PUTTING IT ALL TOGETHER

♀PG-79.   One way of writing a **general equation** for a **parabola** is $y = a(x - h)^2 + k$. This equation tells us how to start with the parent $y = x^2$ and shift or stretch it to get any other parabola.

    a)   Explain what each letter (a, h, & k) represents in relation to the graph of $y = x^2$. Add this information to your Tool Kit.

    b)   Write a general equation for a <u>cubic</u> with parent graph $y = x^3$, and explain how each letter in your general equation affects the stretch or location. Make sure everyone in your group agrees, and be ready to present your ideas to the class.

PG-80.   Sketch a graph of these cubics without a graphing calculator. Then use your calculator to check your thinking.

    a) $f(x) = 3(x - 2)^3 + 2$ b) $y + 3 = \frac{1}{3}(x + 1)^3$

♀PG-81.   Take out your Parent Graph Tool Kit. As a group, write **general equations** for <u>each</u> parent. Be ready to explain how your general equations work; that is, tell what effect each part has on the graph: openness & direction, horizontal location (left/right shift), vertical location (up/down shift). Include an example with a sketch.

PG-82.   Carefully graph $y = x^3 - 4x^2 - 3x + 18$ for $-4 < x < 4$.

a)   How many x-intercepts does it have and what are their coordinates?

b)   What <u>small</u> change you could make in the equation so that the graph would have:

   (i)   one x-intercept.

   (ii)   three x-intercepts.

   (iii) four x-intercepts.

PG-83.   Tasha is experimenting with her graphing calculator. She claims that $y = \frac{1}{x}$ is the parent of $y = \frac{4}{x}$. Do you agree? If not, tell why not. If you do agree, what change (shift, stretch, or compress) could you make to $y = \frac{1}{x}$ to get $y = \frac{4}{x}$?

PG-84.   Explain the difference between the graphs of $y = \frac{1}{x}$ and $y = 4\left(\frac{1}{x + 5}\right) + 7$.

PG-85.   How is $y = 2^x$ different from $y = -(2^x)$? Sketch the graph of $y = -(2^x)$.

PG-86.   Sketch the graph of $y = 2(x - 1)^2 + 4$.

a)   Now rewrite the equation $y = 2(x - 1)^2 + 4$ without parentheses (that is, perform whatever algebra is necessary to get rid of the parentheses. Remember order of operations!).

b)   What would be the difference between the graphs of the two equations in parts (a) and (b) above? This is sort of a trick question, but do explain your reasoning.

c)   What is the parent of $y = 2(x - 1)^2 + 4$?

d)   What is the parent of $y = 2x^2 - 4x + 6$?

PG-87.   Imagine a sphere being inflated so that it gets larger and larger, but remains a perfect sphere. Consider the function defined this way: the inputs are the radius of the sphere, and the outputs are the volume. You may remember from a previous math course that the formula for the volume of a sphere is $V(r) = \frac{4}{3}\pi(r)^3$.

a)   What is the parent of this function? Explain how you decided.

b)   What is the stretch factor?

c)   If the radius is measured in centimeters, what are the units for the volume?

d)   Make a sketch of this function.

e)   What are the domain and range?

PG-88.   The cables that hold up a suspension bridge (like the Golden Gate Bridge) also have the shape of a parabola (surprise!). Find an equation to model the section of cables between the towers of the bridge in the diagram.

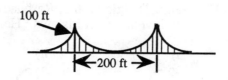

PG-89.   The amount of profit (in millions) made by Scandal Math, a company that writes math problems based on tabloid articles, can be found by the equation: $P(n) = -n^2 + 10n$, where **n** is the number of textbooks sold (also in millions). Find the maximum profit and the number of textbooks that Scandal Math must sell in order to attain this maximum profit.

PG-90.   Draw the graph of: $y = 2x^2 + 3x + 1$.

      a)   Find the x and y intercepts.

      b)   Where is the line of symmetry of this parabola? Write its equation.

      c)   Find the coordinates of the vertex.

PG-91.   Change the equation in the previous problem so that the parabola has only one x intercept.

PG-92.   Shortcut Shuneel claims he has an easier way of changing the equation so that the parabola has only one x intercept.

      He just changed $y = 2x^2 + 3x + 1$ to $y = 0x^2 + 3x + 1$. What do you think of his method?

PG-93.   Consider equations in the form $y = ax^2 + bx + c$, where a, b, and c are constants (numbers).

      a)   Make up three examples of functions that are in this form. What is the parent of each one?

      b)   Is $y = ax^2 + bx + c$ always a parabola no matter what a, b, and c are? Explain fully with examples to support your ideas.

PG-94.   Jamal's little brother Rashaad was watching Jamal make function machines. He decided
          to invent one of his own. Here are four examples of what Rashaad's machine does:

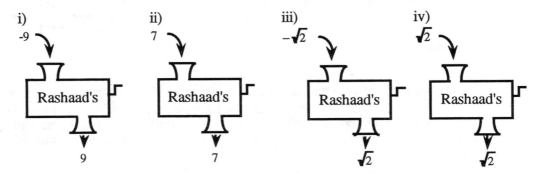

          What is Rashaad's rule? (it's okay just to describe the rule in words).

PG-95.   The mathematical name for Rashaad's machine is ABSOLUTE VALUE, and we write
          the rule for his function using a special symbol: $f(x) = |x|$. With this in mind, find:

          a)   f(-6)   b)   f(13)   c) $f(-\frac{3}{5})$

          d)   f(-π)   e)   f(0)   f)   f(n)

PG-96.   The following problem is a chance for you to demonstrate what you've learned so far
          in this chapter. Try to be accurate and thorough.

          a)   Investigate $y = |x|$. Use a full sheet of graph paper, and be sure to include all
                the elements of a function investigation.

          b)   Take out your Parent Graph Tool Kit. Do you think that any of the equations you
                already have is the parent of $y = |x|$? Why or why not?

          c)   How can we change the equation $y = |x|$ to move it up, down, left, or right?
                How can we stretch it or compress it, or to make it open down? Include several
                examples, and write a general equation for $y = |x|$.

          d)   What questions do you still have? If you can't think of any, come up with a good
                extension question--that is, what would be a good follow-up question to explore?

PG-97.   Add $y = |x|$ to your Parent Graph Tool Kit. Include everything you found in your
          investigations.

PG-98.   For what values of x does:

          a) $|x| = 8$?   b) $|x| = 4$?

          c) $|x| = 0$?   d) $|x| = -4$?

PG-99.   Olivia's group was working on the problem: $|x + 1| = 6$. The other three people in her group all got $x = 5$ as their answer, but Olivia got a different answer. Her group-mates all tried to convince Olivia that her answer was wrong, but Olivia knew that it was correct because she checked it. She said that <u>both</u> answers were right.

a)   What was Olivia's answer?

b)   Explain why both answers are correct and show how you could get both solutions.

c)   Make a sketch of $f(x) = |x + 1|$ and use the graph to show why there are two answers. Identify on your graph what two x-values give a corresponding y-value of 6.

PG-100.   Graph $f(x) = |x - 3|$ and $x = 3$ on the same set of coordinate axes. Write a description of the relationship between the two graphs.

PG-101.   Compare these two expressions: $|11 - 5|$ and $|5 - 11|$.

a)   How would you simplify each of them?

b)   Explain why you get the same answer.

PG-102.   For what values of x does:

a) $|x - 7| = 50$?

b) $|x + 7| = 50$?

c) $|10 - x| = 12$?

d) $|2x + 1| = -3$?

PG-103.   Explain the difference between $f(x) = |x|$ and $y = |x|$.

PG-104.   Sketch a graph each of the following equations and draw the line of symmetry for each.

a) $y = |x - 4|$

b) $g(x) = -|x + 1|$

c) $f(x) = |x + 3| + 6$

d) $y + 3 = 2|x - 4|$

e)   Write the equations for the lines of symmetry you found in parts (a) through (d) above.

PG-105.   Find the intercepts, the locator point, and give the domain and the range for each of the following functions:

a) $y = |x - 4| - 2$

b) $y = -|x + 1| + 3$

PG-106.   By mistake Jim graphed $y = x^3 - 4x$ instead of $y = x^3 - 4x + 6$. What should he do to his graph to get the correct one?

PG-107. If $x^2 + kx + 18$ is factorable, what are the possible values for k?

PG-108. Consider this system of equations: $3y - 4x = -1$
                                                $9y + 2x = 4$

 a) What is the parent of each equation?

 b) Solve this system.

 c) Find where the two lines intersect.

 d) Explain the relationship between parts (b) and (c) above.

PG-109. Write the description for a situation that could fit the following function:

 $$f(t) = 2000(0.91)^t$$

PG-110 Write an equivalent expression without parentheses for each of the following.

 a) $(2x^2y)^3 =$ b) $(5x + 0.5)^2 =$

 c) $5(2s - 7)(2s + 7) =$ d) $(5d^{-5})^3(-3d^3)^5 =$

TWO NON-FUNCTIONS

PG-111. If you haven't already done so, try to find a way to graph $x = y^2$ on your graphing calculator. Since your graphing calculator only graphs functions in y-form, you will have to be clever. Talk to your group and come up with a way of getting $x = y^2$ to graph on your calculator.

PG-112. Now try to graph $x^2 + y^2 = 9$ on your graphing calculator. What equation(s) will you use?

PG-113. Investigate $x^2 + y^2 = 9$. Along with all the usual elements of an investigation, discuss also whether this equation is or is not a function.

PG-114. What would be the equation of a circle with radius 10? With radius 12? With radius r?

PG-115. Where is the center of $x^2 + y^2 = 9$? How do you think you could change the equation so that the center of the circle is moved to $(4, 0)$? $(0, 4)$? $(3, -5)$? to (h, k)?

PG-116. Rewrite the last equation from the preceding problem so the radius could be any length. You now have a **general equation** for a **circle** from which you can read its radius and locator point (otherwise known as its center). Add a circle and all of the associated information to your Parent Graph Tool Kit.

PG-117. Think about the equation $x = y^2$. How is $x = y^2$ different from all except the equation for the circle?

    a)   Investigate $x = y^2$.

    b)   How would you change the equation $x = y^2$ to:

        (i)   move its graph up 2 units?

        (ii)   move its graph 6 units to the right?

        (iii)   make its graph open wider(shrunk)?

        (iv)   make its graph open narrower(stretched)?

        (v)   make its graph open to the left?

    c)   Can you use your graphing calculator to check your work? Explain.

PG-118. Use your ideas from the previous problem to write a general equation for all parabolas that open sideways ("sleeping parabolas"). It is <u>not</u> necessary to add this one to your Tool Kit.

PG-119. Remember Rashaad's "Absolute Value" machine? Take a look at it again. Whenever Rashaad plugged a number into his machine, he got <u>only one</u> number out. Whenever a relationship has this property (for each input there is one and only one output), the relationship is called a **FUNCTION**.

    a)   Sketch a graph of $y = x^2$. Is it a function? Explain how you know.

    b)   Is $x = y^2$ a function? Why? Go back to your investigation and add a statement about whether or not $x = y^2$ is a function, and what that means.

    c)   Add the definition of a function to your Tool Kit, and include an example of a graph that is a function and one that is not.

PG-120. Curtis, who lives on a farm and likes cows, has invented a weird type of machine, which he calls *"Curtis' Cow Machine."* **Each picture represents the same machine.

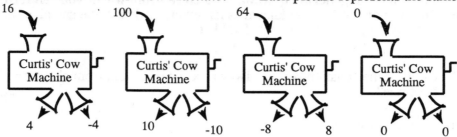

a) Is this machine a <u>function</u> machine? Explain.

b) Draw a graph that represents his machine. Remember that the numbers on top are the inputs and the numbers that come out of the cow machine are the outputs.

c) Write a rule for Curtis's Cow Machine.

PG-121. Anne says that $y = x^2$ is not a function, because when she plugs in 3 for x, she gets 9, and when she plugs in -3 she gets 9 too. Two different inputs give her the same output. Brooke says that Anne has the function idea all mixed up. Help! Explain who's right. Is $y = x^2$ a function or not? Why?

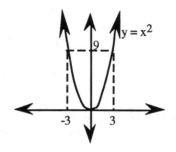

PG-122. Of the parents listed in your **Parent Graph Tool Kit**, which are functions? Explain.

PG-123. Find the x and y-intercepts and the locator points of:

a) $y - 7 = 2x^2 + 4x - 5$

b) $x^2 = 2x + x(2x - 4) + y$

PG-124. For each of the following, write an equation and sketch a graph for the circle with:

a) locator point: $(0, 0)$, radius: 5.

b) locator point: $(2, 3)$, radius: 5

c) locator point: $(-1, -4)$, radius: 5

PG-125. A parabola has vertex $(2, 3)$ and contains $(0, 0)$. Find its equation.

PG-126.   Solve for **x**.

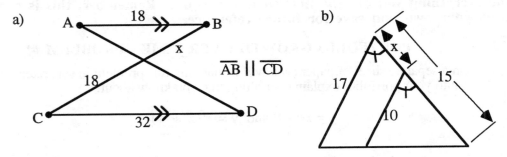

a) A — 18 — B, x, 18, C — 32 — D, $\overline{AB} \parallel \overline{CD}$

b) triangle with 17, 15, 10, x

PG-127.   The area of a square is 225 sq. cm.

a)   Make a diagram and list any subproblems that you would need to do to find the length of the diagonal.

b)   What is the length of its diagonal?

THE GATEWAY ARCH (REVISITED)

PG-128.   Remember why we wanted to move parabolas around in the first place? We wanted the specific equation for a parabola that could approximate the Gateway Arch. Now we can do it! Go back and read problem PG-1 again, and create an equation. Write out your work in good detail, explaining your ideas where necessary.

CHAPTER 4 SUMMARY ASSIGNMENT
(problems PG-129-131)

PG-129.   Write an explanation relating a graph to its parent graph, so that someone new to this class could understand how to tell whether the stretch factor of its equation was:

a)   positive or negative.

b)   less than 1 or greater than 1.

PG-130.   Each of the general equations in your Parent Graph Tool Kit has a **locator point**. For each parent:

a)   describe what about the graph the locator point locates.

b)   Write an equation for each parent with locator point (3, -4) and sketch each graph.

PG-131.   Look back over your work on the labs and investigations that you did in this chapter. Write several paragraphs summarizing what you've discovered. What new mathematics did you learn? What were the key ideas? What conclusions, conjectures and generalizations can you make? What unanswered questions do you still have?

This is the second time around for this growth over time problem. Be sure you include everything you did the first time and more. Remember, this is a problem to do on your own and save for future reference.

EF-132.   **PORTFOLIO GROWTH OVER TIME - PROBLEM #1**

On a separate sheet of paper (you will be handing this problem in separately or putting it into your portfolio) explain everything that you know about:

$$y = x^2 - 4 \quad \text{and} \quad y = \sqrt{(x + 4)}.$$

PG-133.   For each of the following equations, explain what **d** does to the shape, size and/or location of the equations' graphs:

a) $y = d \, | \, x \, |$

b) $y = 3x^2 - d$

c) $y = (x - d)^2 + 7$

d) $y = \dfrac{1}{x} + d$

PG-134.   Sketch a graph of $g(x) = x^2 - 2x$.

a)   Is it a function? How do you know?

b)   What are the coordinates of its vertex?

c)   What are its domain and range?

PG-135.   Given: $f(x) = x^3 + 1$ and $g(x) = (x + 1)^2$,

a)   sketch the graphs of the two functions.

b)   find: $f(3)$.

c)   solve: $f(x) = 9$.

d)   find: $g(0)$.

e)   solve: $g(x) = 0$.

f)   solve: $f(x) = -12$.

g)   solve: $g(x) = -12$.

PG-136.   Find the intercepts (both x and y), inflection points, and vertex of each function in the previous problem. Then find the equations of any lines of symmetry.

PG-137.   Write an equation for this graph, assuming that it has y-intercept (0, 1):

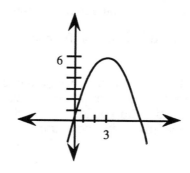

The following problem is somewhat open-ended. There is not enough information to get the precise equation of the parabola, so you will need to draw a reasonable trajectory and estimate--that is, model a possible trajectory with an equation.

PG-138.   An archer shoots an arrow at a target that is 40 meters away. If the arrow strikes the target at the same height above the ground as it was released(about 1.3m above the ground or at "eye level"), choose a reasonable maximum height for the flight of the arrow and find a possible equation to model the path based on the maximum that you chose. Results will vary depending on what people choose for a reasonable maximum height.

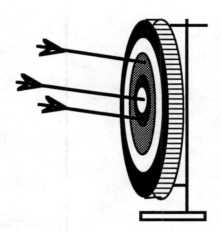

PG-139.   Sketch a graph of:

a) $g(x) = (x + 3)^3$ b) $y - 1 = 3^x$ c) $(x + 2)^2 + (y - 1)^2 = 25$

PG-140.   If $p(x) = x^2 + 5x - 6$, find:

a)   where $p(x)$ intersects the y-axis,

b)   where $p(x)$ intersects the x-axis.

c)   Now suppose $q(x) = x^2 + 5x$. Find the intercepts of $q(x)$ and compare the graphs of $p(x)$ and $q(x)$.

PARENT GRAPH TOOL KIT

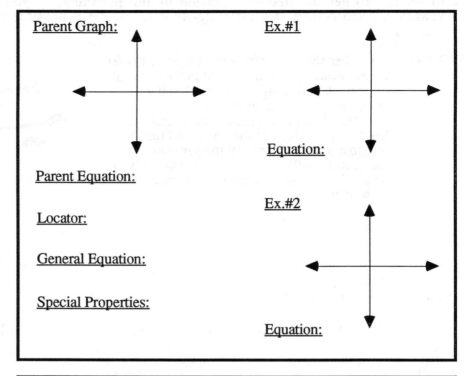

_____:

Parent Graph:

Parent Equation:

Locator:

General Equation:

Special Properties:

Ex.#1

Equation:

Ex.#2

Equation:

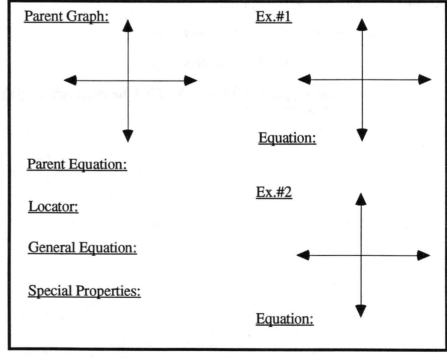

_____:

Parent Graph:

Parent Equation:

Locator:

General Equation:

Special Properties:

Ex.#1

Equation:

Ex.#2

Equation:

Chapter 5
Linear Systems
(The Toy Factory)

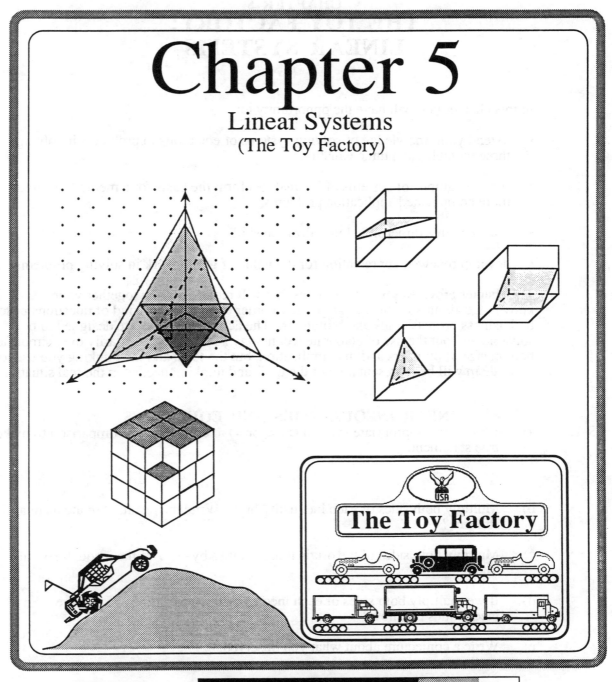

The Toy Factory

PROBLEM SOLVING		
REPRESENTATION/MODELING		
FUNCTIONS/GRAPHING		
INTERSECTIONS/SYSTEMS		
ALGORITHMS		
REASONING/COMMUNICATION		

CHAPTER 5
THE TOY FACTORY:
LINEAR SYSTEMS

In this chapter you will have the opportunity to:

- extend your knowledge of solving systems of equations to problems involving three variables and three equations.

- see how graphs of systems of inequalities form the basis for a method to solve more complicated application problems.

- practice three dimensional visualization skills.

- develop renewed appreciation for the value of group work in solving problems.

The chapter provides several problems that will require pulling together your knowledge about solving and graphing in order to address the kind of questions posed by business professionals and biologists. The actual questions are designed to be solvable without the use of elaborate technology so they are necessarily over simplified both in size of numbers and in complication, but the tools and the thinking you use to solve them will be representative of the tools and thinking needed in the real situation.

LINEAR INEQUALITIES AND EQUATIONS

LS-1.   a)   Choose the appropriate symbol (>, <, or =) to complete the comparison to make a true statement.

   i) -2 $\underline{?}$ 3   ii) 6 $\underline{?}$ 4   iii) -5 $\underline{?}$ -1

   b)   Multiply both sides of each inequality by 5. Write the result. Are the inequalities still true?

   c)   Multiply both sides of the original inequalities by -1. Are the inequalities still true?

   d)   If we multiply both sides of each inequality from part (a) by -5, are the inequalities true now?

   e)   Write a conjecture about what you observed.

LS-2.   In the previous problem you determined that multiplying both sides of an inequality by a negative number makes the inequality false. What can you do to the inequality sign to make the inequality true again?

♀LS-3.   The conjectures you made in the previous problems will help you transform inequalities to y-form accurately.

   a)   Choose a value of **y** that makes the inequality $-2y < 6$ obviously true.

   b)   Write the inequality in y-form (that is, solve for y). Test the value you choose in (a) on your transformed inequality. Is it still true? What must you do to make it true again?

   c)   Repeat this test on the following inequalities. Choose a value that makes the inequality true, write the inequality in y-form, then test the value again to be sure that you have adjusted the inequality correctly to keep it true.

      i) $-3y + 1 > 4$

      ii) $2y + 7 > 5$

      iii) $6 - 4y < 0$

   d)   Write a rule about multiplying and dividing inequalities by negative numbers. Be sure to include this rule and several examples in your Tool Kit.

LS-4.   When solving an inequality, what causes the inequality symbol to reverse its order?

LS-5.   Remember when drawing the graph of $y > 3x - 1$, we begin by graphing the line $y = 3x - 1$ with a dashed line. People decided to use a dashed line instead of a solid line for problems like this one. Do you think that was a good idea? Why would it be important to use something different from a regular line?

Point	Location (Above, Below, or On)	$y > 3x - 1$ (True or False)
A(0, 0)		
B(3, 4)		
C(-1, -4)		
D(-2, 3)		
E($\frac{-2}{3}$, -4)		

   a)   Graph the line (dashed) and plot the points from the table with your line. Label each point.

   b)   Complete the table. Which of these points makes $y > 3x - 1$ true?

   c)   Which region of your graph should be shaded to represent all the points for which $y > 3x - 1$ is true? Describe the region <u>and</u> shade it.

LS-6.   Simone has been absent and does not know the difference between the graph of $y \leq 2x - 2$ and the graph of $y < 2x - 2$. Explain thoroughly so that she completely understands what points are excluded from the second graph.

LS-7.   Graph the solution to each of the following inequalities on a different set of axes (but you should be able to fit all four on one side of the graph paper). Label each graph with the inequality as given <u>and</u> with its y-form. Choose a test point and show that it gives the same result in both forms of your inequality.

a) $3x - 3 < y$ b) $3 > y$

c) $3x - 2y \leq 6$ d) $x^2 - y \leq 9$

LS-8.   Since she had been doing so well in her algebra class, Lorrel decided to reward herself with her own phone line (her roommate is very talkative). Lorrel's boyfriend lives in the Bay Area and she needs to choose a long distance plan. The phone company offers two plans: Normal Rate, which will cost $12 for each hour of use, or High Use Rate which has a monthly fee of $12 plus $10 for each hour of use.

a)   Write an equation, one for each plan, which will represent Lorrel's long distance cost based on the number of hours she calls her boyfriend.

b)   Graph each of the equations you wrote in part (a) on the same set of axes. Under what conditions is it better for Lorrel to use the Normal Rate? When is it better to use the High Use Rate?

c)   When could either of the company's plans be used for the same price? What is the significance of this value? How does it relate to the graph?

LS-9.   Angela and Zack are going to a Rolling Stones concert. They agreed that Angela would pay $105 for the tickets. Zack would pay for the limo rental. A Cadillac rents for $50 plus $6 per hour. A Lincoln rents for $25 plus $12 per hour. Write an equation to represent each company's cost using hours as the independent variable. Graph both on the same set of axes. Be sure to label the axes.

a)   For what number of hours can Zack rent either car for about the same cost? (Find your answer to the nearest $1/2$ hour.) How much is the cost?

b)   If Zack rents the car for 7 hours, which car should he use? How much is the cost?

c)   Zack wants to spend the same amount as Angela. Which car should he rent to get the most time? How long will they have the car?

LS-10.   Sketch and describe the graphs of:

a) $y = 3$ b) $x = -2$ c)   Where do they cross?

LS-11.   a)   Graph the following on the same set of axes:

i) $y = 3$ ii) $y = 2$ iii) $y = 1$

b)   What would be the equation of the x-axis?

c)   What would be the equation of the y-axis?

LS-12.   Solve this system of equations:

$$2x + 6y = 10$$
$$x = 8 - 3y$$

a)   Describe what happened.

b)   Draw the graph of the system.

c)   How does the graph of the system explain what happened with the equations? Be clear.

d)   What might be a similar situation in three dimensions?

LS-13.   Kamau needs to find the point of intersection for the lines $y = 18x - 30$ and $y = -22x + 50$. He takes out a piece of graph paper and then realizes that he can solve this problem without graphing. Explain how Kamau is going to accomplish this and find the intersection point.

LS-14.   Find the x and y intercepts.

a) $2x - 3y = 9$ b) $3y = 2x + 12$

LS-15.   Graph each of the following. (Keep these handy, you will need them later).

a) $y = |x|$ b) $|y| = x$

c)   How are the two graphs similar or different?

d)   What is the domain and range of each?

LS-16.   Graph the inequalities below on the same set of axes. Shade the solution of each.

$$y \geq -4x - 2$$
$$3x + 2y < 6$$

a)   Test the following points in both inequalities and label them on your graph. Indicate which ones are solutions to both inequalities.

A(0, 4)   B(0, 0)   C(-1, -1)   D(4,3)

b)   What does the area where the graphs overlap represent?

SYSTEMS OF INEQUALITIES

LS-17.   Take out your graph of the overlapping inequalities from the homework. Discuss your solutions with your group, and answer the following:

a)   When do we use a dashed line instead of a solid line?

b)   What determines where the final region is shaded?

LS-18.   Graph the four inequalities below on the same set of axes. Before you begin it would
          be a good idea for the group to discuss the most efficient way to graph a line. Be very
          careful to compare your answers as you work through the parts!

   i) $2y \geq x - 3$ ii) $x - 2y \geq -7$

   iii) $y \leq -2x + 6$ iv) $-9 \leq 2x + y$

   a)   What type of polygon is formed by the solution of this set of inequalities? Write a
          convincing argument to justify your answer.

   b)   Find the vertices of the polygon. If your graph is very accurately drawn you will
          be able to determine the points from the graph. If it is not, you will need to solve
          the systems (pairs) of equations that represent the edges of your graphs.

LS-19.   Find the area of the polygon that you graphed in the last problem.

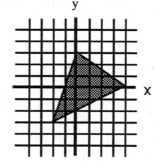

LS-20.   Write the three inequalities that will form the triangle
          shown at right.

LS-21.   Use these three equations to solve for **x**, **y** and **z**.

$$3x + 8 = 2$$
$$7x + 3y = 1$$
$$\frac{1}{2}x + y - 8z = 8$$

LS-22.   The Alvarez family plans to buy a new air conditioner. They can buy the Super Cool
          X140 for $800 or they can buy the Efficient Energy 2000 for $1200. Both models will
          cool their home equally well, but the Efficient Energy model is less expensive to
          operate. The Super Cool X140 will cost $60 a month to operate while the Efficient
          Energy 2000 costs only $40 a month to operate.

   a)   Write an equation that will represent the cost of buying and operating the Super
          Cool X140.

   b)   Write an equation that will represent the cost of buying and operating the Efficient
          Energy 2000.

   c)   How many months would the Alvarez's have to use the Efficient Energy model to
          compensate for the additional cost of the original purchase?

   d)   Figuring they will only use it 4 months each year, how many years will they have
          to wait to start saving money overall?

LS-23.   Macario's salary is increasing by 5% each year. His rent is increasing by 8% each year. Currently, 20% of Macario's salary goes to pay his rent. Assuming that Macario does not move or change jobs, what percent of his income will go to pay rent in 10 years? Think about this problem and if you see a way to solve it, do it. If not try the a), b), c) and d) parts.

a)   If **x** represents Macario's current salary write an expression to represent his salary ten years from now.

b)   Use **x** in an expression to represent Macario's rent now.

c)   Use what you wrote in part (b) to write an expression for the rent ten years from now.

d)   Now use the expressions that you wrote in parts (a) and (c) to write a ratio that will help you answer the question.

LS-24.   It is often useful to use isometric dot paper to help visualize solids in two dimensions.

This is an **ISOMETRIC DOT PAPER GRID**. IMPORTANT DETAIL: Isometric dot paper is not the same if it is turned sideways. To be sure that your paper is correctly lined up, draw a hexagon like the one shown below. If the sides are vertical, then the paper is turned correctly; if the sides have points, then turn the paper.

Another way to check that your paper is turned correctly is to look at the left margin of the paper. The dots should make a straight, vertical column. If the dots zigzag in and out between rows, the paper is turned the wrong way.

This is turned correctly.   This is not!

This is a **3-D ISOMETRIC** drawing of a cube.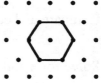

By stacking and pushing cubes together, we can create other solids. Two cubes can be joined to form two isometric solids as shown at right:

Keep real cubes handy as you are working on these problems. Setting up what is pictured using real cubes can be very helpful when interpreting the drawings. When we look at an isometric drawing we are ALWAYS viewing it from the front, right, and top parts of the solid.

On isometric dot paper make a drawing that includes 9 cubes with at least one of them hidden.

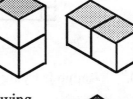

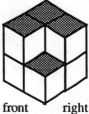

front   right

LS-25. Each cube is 1 cm on a side.

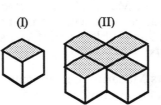
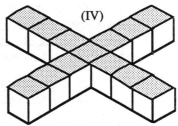

a) Based on the pattern, draw figure (III) on isometric dot paper.

b) Find the volume of each figure (I) through (IV).

c) The pattern continues. Write an expression to represent figure N. What kind of sequence is this?

LS-26. Rewrite as an equivalent expression without negative exponents

a) 5^{-2}

b) xy^{-2}

c) $(xy)^{-2}$

d) $a^3b^4a^{-4}b^6$

LS-27. If a 3 cm. cube is first painted red and then cut into twenty-seven 1 cm. cubes. All twenty-seven 1 cm cubes are placed in a bag and one is drawn out at random. What is the probability that the cube has paint on

a) 0 sides?

b) 1 side?

c) 2 sides?

d) 3 sides?

e) 4 sides?

LS-28. Compare and contrast the graphs of $y > x^2$ and $y > (x - 4)^2 + 2$.

LS-29. Solve for x and y in the following problems.

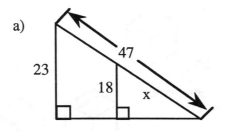

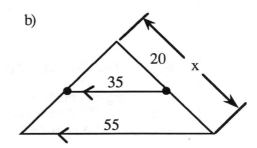

LS-30. Ramin is trying to evaluate an expression and he can not get the negative sign to work on his calculator! Explain to Ramin how he can simplify $\dfrac{7 \cdot 4^{-2001}}{2 \cdot 4^{-1997}}$ without using a calculator.

POINTS IN SPACE

LS-31.   Draw and label a three-dimensional coordinate axis system like the one at right on isometric paper. Discuss how the drawing is related to the origami structure you have.

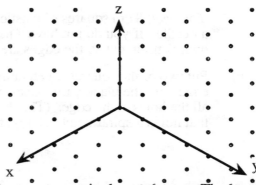

LS-32.   When locating a point on a number line, we use a single number, **x**. The location of a point on a plane is given by two numbers, (x, y), called an ordered pair. To locate a point in space, we use three numbers, (x, y, z), called an **ordered triple**. Put a cube in your origami structure so that it is in the first octant and has (0, 0, 0) as a vertex.

   a)   Draw the cube sitting in the octant on isometric dot paper.

   b)   Suppose that the length of an edge of the cube is 3. Find the coordinates of the other seven vertices.

LS-33.   A rectangular prism is placed in your origami structure so that one vertex is at the origin and the vertex furthest from the origin is at the point (3, 2, 5).

   a)   Draw a picture of this prism in the first octant. Use isometric dot paper.

   b)   Find the coordinates of the other seven vertices.

   c)   What is the volume of the prism?

LS-34.   Points A, B, and C are on the axes shown at right.

   a)   Find the coordinates of A, B, and C on the graph.

   b)   Teller K. Draper plotted the point D. The path he took to get to D is shown. What are the coordinates of point D?

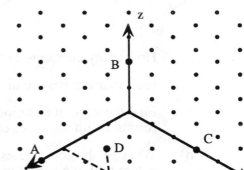

LS-35.   Refer back to the graphs you did on LS-15. Use those graphs to help you to graph each of the following.

   a) $y \leq |x|$ b) $|y| \geq x$

LS-36.   In this activity you will be making a set of axes to represent the first octant of the 3-D axes. We will be using this as a model a few days from now.

   a)   You need 8 cm squares of plastic netting, approximately 45 cm of black yarn, and a needle. If you do not have a needle, wrap tape around the end. If you cut your own netting, cut so the edges are as smooth as possible.

   b)   Put two of the squares together at right angles and sew the edge with the black yarn. Use a whip stitch and go through all the holes at the edge. (The threading is similar to the wire that holds a spiral notebook together.)

   c)   Place the third square to form a corner of a cube. Sew all of the edges in black. Your result will be one **octant** of the 3-D axes.

LS-37.   Solve the following system algebraically and explain what the solution tells you about the graphs of the two equations.

$$4x - 6y = 12$$
$$-2x + 3y = 7$$

LS-38.   Use isometric dot paper to draw the figure for n = 4.

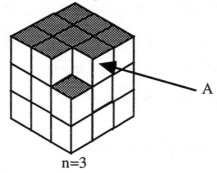

A

n=2   n=3

   a)   How many cubes are in the figure for n = 5?

   b)   How many cubes are in the figure for n = 1?

   c)   Find the general equation for the number of cubes for any given n. Verify your formula with the cases of n = 1 and n = 5.

   d)   What are the coordinates of point A when the figure for n = 3 is placed in the first octant with the vertex we can't see at (0, 0, 0)?

LS-39.   Marvelous Mark's Function Machines. Mark has set up a series of three function
          machines that he claims will surprise you:

          a)   Try a few numbers. Were you surprised?

          b)   Carrie claims she was not surprised and
                she can show why the sequence of
                machines does what it does by simply
                dropping in a variable and writing out
                step-by-step what happens inside each
                machine. Try it (use something like **c** or
                m). Be sure to show all the steps.

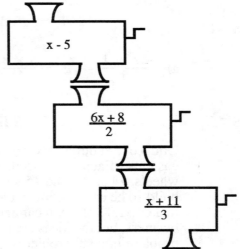

LS-40.   Hamida inadvertently conducted an experiment by leaving her bologna sandwich at
          school over the winter break. Much to her surprise, her sandwich (or what used to be a
          sandwich) was much larger than it was when she left it.

          After inspecting it, her science teacher said Hamida had produced great quantities of a
          seldom seen bacteria, bolognicus sandwichae. Based on a sample taken from the
          sandwich, they determined that there were approximately 72 million bacteria present.
          Hamida was greatly surprised by this number. Her biology teacher informed her that
          this is not too surprising since the number of bacteria triples every 24 hours. Hamida
          thought that the sandwich must have been loaded with bacteria initially since it had been
          made only 15 days ago.

          Hamida is making plans to sue the meat company because she learned that the food
          industry standard for the most bacteria a sandwich could have at the time it was made
          was 100. Find out how many of the bacteria were present when the sandwich was
          made to determine if Hamida has a case.

LS-41.   Rewrite as an equivalent expression using fractional exponents

          a) $\sqrt[2]{5} =$ b) $\sqrt[3]{9} =$ c) $\sqrt[8]{17^x} =$

LS-42.   Draw the graph of this system of inequalities:

                $y \geq |x| - 3$
                $y \leq -|x| + 5$

          a)   What polygon does the intersection form? Justify your answers.

          b)   What are its vertices?

          c)   Find the area of the intersection.

LS-43.   Solve for x: $2^x = 3$. Be accurate to three decimal places.

LS-44.   Solve:

a) $\dfrac{2z + 1}{5} - \dfrac{z}{10} = \dfrac{1 - z}{4}$ b) $\dfrac{2w + 1}{3w} = \dfrac{2w}{3w - 1}$

LS-45.   **THE TOY FACTORY**

Otto Toyom builds toy cars and toy trucks. Each car needs 4 wheels, 2 seats, and 1 gas tank. Each truck needs 6 wheels, 1 seat, and 3 gas tanks. His storeroom has 36 wheels, 14 seats, and 15 gas tanks. He needs to decide how many cars and trucks to build so he can maximize the amount of money he makes when he sells them. He makes $1.00 on each car and $1.00 on each truck he sells. (This is a very, very small business, but the ideas are similar for much larger enterprises.) We will divide this problem into subproblems.

a)   Otto's first task is to figure out what his options are. For example, he could decide to make no cars and no trucks and just keep his supplies. On the other hand, because he likes to make trucks better, he may be thinking about making five trucks and one car. Would this be possible? Why? What are all the possible numbers of cars and trucks he can build, given his limited supplies? This will be quite a long list. An easy way to keep your list organized and find some patterns is to plot the points that represent the pairs of numbers in your list directly on graph paper. Use the x-axis for cars and the y-axis for trucks. Make a fairly large, neat first quadrant graph. (Why the first quadrant?) We will need to use this graph later in this problem and in the next.

b)   In part (a) you figured out all the possible combinations of numbers of cars and trucks Otto could make. Which of these give him the greatest profit? Explain how you know your answer is right. You have to convince Otto, who likes trucks better.

c)   New scenario: Truck drivers have just become popular because of a new TV series called "Big Red Ed." Toy trucks are a hot item. Otto can now make $2.00 per truck though he still gets $1.00 per car. He has hired you as a consultant to advise him, and your salary is a percentage of the total profits. What is his best choice for the number of cars and the number of trucks to make now? How can you be sure? Explain.

LS-46.   In the last problem you probably had to carry out a lot of calculations in order to
          convince Otto that your recommendation was correct. In this problem we will take
          another look at Otto's business using some algebra and graphing tools we already
          know.

     a)   The first task is to write three inequalities to represent the relationship between the
          number of cars, x, the number of trucks, y and the number of:

          i)   wheels.   ii)   seats.   iii)   gas tanks.

     b)   Carefully graph this system of inequalities (assume $x \geq 0$ and $y \geq 0$, why?) on the
          same set of axes you used for the last problem. Shade the region of intersection
          lightly.

     c)   What are the vertices of the pentagon that outlines your region? Explain and
          show exactly how and why you could use the five equations below to find those
          five points. In many problems these points are difficult to determine from the
          graph alone.

                    $x = 0$ $2x + y = 14$
                    $y = 0$ $x + 3y = 15$
                    $4x + 6y = 36$

     d)   Part (c) shows how we could get an outline of the region of points that represents
          possible numbers of cars and trucks. Are there some points in the region that
          seem more likely to give the maximum profit? Where are they? Why do you
          think they are the best coordinates? How can you represent the total profit if Otto
          makes $1.00 on each car and $2.00 on each truck?

     e)   What if Otto ended up with a profit of $8? Write an equation for the profit when
          it is $8.00. Use the graph of just one of your group members and draw the graph
          of this equation on it. Do you think from looking at this graph that $8 is the
          maximum Otto could make? Why or why not? Try some other possibilities ($9,
          $10, $11, $12). Write an equation and draw a graph for each (continue to use
          just one person's paper.) Find the maximum and justify your answer for Otto.

     f)   What did all the lines you drew have in common? What was different? What
          were you trying to do? Ask your teacher for a transparency, place it on top of
          some graph paper and draw just a y-axis and one line with the same slope as your
          profit lines. Now put the transparency on top of one of your group's graphs and
          align the y-axis. How could you physically slide the transparency over the graph
          to find the maximum profit? Try it on another group member's graph and explain
          why this method should work.

LS-47.   Use the method you developed in the last problem to find Otto's maximum possible
          profit if he gets $3.00 per car and $2.00 per truck.

LS-48.   Graph the inequalities and calculate the area bounded by them.

                    $y \leq 2x + 6$
                    $y \leq -x + 3$
                    $y \geq -2$

LS-49.   Solve the following system:

$$2^{x+y} = 16$$
$$2^{2x+y} = \frac{1}{8}$$

LS-50.   Anywhere High School has an annual growth rate of 4.7%. Three years ago there
           were 1500 students.

           a)   How many students are there now?

           b)   How many 5 years ago?

           c)   How many in **n** years?

LS-51.   Paul states that $(a + b)^2$ is equivalent to $a^2 + b^2$. Joyce thinks that Paul is incorrect with
           his statement. Help Joyce show Paul that the two expressions are not equivalent.

LS-52.   Factor each of the following:

           a)   bx + ax   b)   x + ax

           Factor and reduce each of these:

           c) $\dfrac{ax + a}{x^2 + 2x + 1}$ d) $\dfrac{x^2 - b^2}{ax + ab}$

LS-53.   From each of the following pairs of domain and range "shadow" graphs, sketch two
           different graphs that satisfy the conditions of each pair of shadows.

           a)   b)

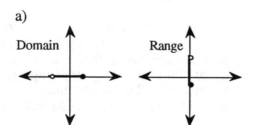   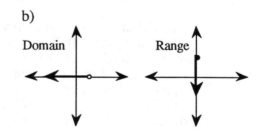

LS-54.   Sketch each graph:

           a) $(x - 2)^2 + (y + 3)^2 = 9$ b) $(x - 2)^2 + (y + 3)^2 \geq 9$

LS-55.   You are standing 60 feet away from a five story building in Los Angeles, looking up at its roof top. In the distance you can see the billboard on top of your hotel but the building is completely obscured by the one in front of you. If your hotel is 32 stories tall and the average story is 10 feet, how far away from your hotel are you?

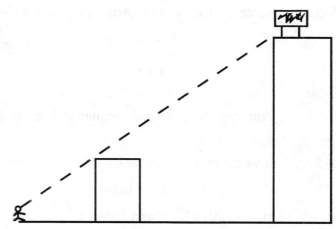

LS-56.   **SANDY DANDY DUNE BUGGIES, INC.**

The Sandy Dandy Dune Buggy Company makes two models of off-road vehicles: the *Sand Crab* and the *Surf Mobile*. They can get the basic parts to produce as many as 15 Sand Crabs and 12 Surf Mobiles per week, but two parts have to be special ordered: a unique exhaust manifold clamp and a specially designed suspension joint. Each Sand Crab requires 5 manifold clamps and 2 suspension joints. Each Surf Mobile requires 3 manifold clamps and 6 suspensions joints. For each week the maximum number of clamps available is 81 per week and the maximum number of joints is 78 per week.

It takes 20 hours to assemble one Sand Crab and 30 hours to assemble one Surf Mobile. The company has 12 employees who each work a maximum of 37.5 hours per week.

Sandy Dandy, Inc. has hired your group as consultants to advise them on how many Sand Crabs and how many Surf Mobiles they should make each week in order to maximize their profit. They know their profit margin on each Sand Crab is $500 and on each Surf Mobile $1000. Prepare a complete report including your graphs and an explanation to justify your conclusions.

Our goal in this problem is not simply to solve the problem, but to see how it might be solved by using a particular method, **Linear Programming**, that relies on you using your algebraic and graphing skills.

You may choose to work on this problem without further suggestions or you can use the following as a guide.

a)   First you need to use the information in the problem to decide how many combinations of numbers of vehicles can be built. You could do this point by point but it probably is more efficient to write five inequalities and graph them. Work in groups of four but each pair should make a large, neat graph with equations written along the lines they represent.

Let x represent the number of Sand Crabs built in one week.
Let y represent the number of Surf Mobiles built in one week.

b)   You can assume x ≥ 0 and y ≥ 0. Why? What kind of polygon did you get? What would be useful to know about this polygon? Find them.

c)   What expression can you write for the profit? How can you use this expression and your graph to find the maximum possible profit and the number of Sand Crabs and Surf Mobiles that will produce it? Do it, and complete your report.

LS-57.   Solve the following system of equations. What subproblems did you need to solve?

$$x + 2y = 4$$
$$2x - y = -7$$
$$x + y + z = -4$$

LS-58.   Consider the arithmetic sequence 2, a - b, a + b, 35, ... Find the value of a and b.

LS-59.   Give the equation of each circle.

a)   center (0, 0) and radius 6

b)   center (2, -3) and radius 6

LS-60.   A square target 20 cm on a side contains 50 non-overlapping circles each 1 cm in diameter. Find the probability that a dart which hits the board at random hits one of the circles.

LS-61.   Think about the axis system in the two-dimensional coordinate plane. What is the equation of the x-axis? What is the equation of the y-axis?

LS-62.   A line intersects the graph of $y = x^2$ twice, at points whose x-coordinates are -4 and 2.

Draw a sketch of both graphs, and find the equation of the line.

LS-63.   Write each number as a power of 2

a)   16   b) $\frac{1}{8}$

c) $\sqrt{2}$ d) $\sqrt[3]{4}$

LS-64.   An investment counselor advises a client that a safe plan is to invest 30% in bonds and 70% in a low risk stock. The bonds currently have an interest rate of 7% and the stock has a dividend rate of 9%. The client plans to invest a total of x dollars.

a)   Write an expression for the annual income that will come from the bond investment.

b)   Write an expression for the annual income that will come from the stock investment.

c)   Write an equation and solve it to find out how much does the client need to invest to have an annual income of $5,000?

LS-65.   A circle has the equation $x^2 + (y + 2)^2 = r^2$. If the circle is shifted 2 units to the left, 5 units up and the radius is doubled, what will be the new equation.

LS-66.   BE SURE TO BRING YOUR 3-D OCTANT MODEL TO CLASS TOMORROW! (This is the model that you stitched together).

3 - D MODELS AND PLANES IN SPACE

LS-67.   Using the first octant 3-D model you made in LS-35:

a)   Describe which surface represents the xy-plane with respect to the z-axis, and label it on the model. Do the same for the xz-plane and the yz-plane.

b)   Complete at least one of the models below (also on display in the classroom). Use a different color for each model.

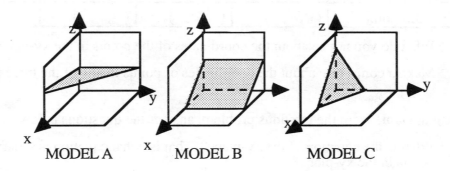

MODEL A   MODEL B   MODEL C

LS-68.   In your groups, use the first octant models you just made to answer the following questions about the first octant 3D model intersecting planes.

a)   Look at Model A, in the previous problem, as a group

i)   Is the colored plane parallel to any coordinate planes? Which one(s)?

ii)   Does it intersect any of the axes? Which one(s)?

iii)   Write the approximate coordinates of the intercepts. Think of the plastic netting as graph paper. Assign your own scale, but be consistent for the three models.

iv)   What is the equation of the plane?

b)   Look at Model B as a group.

i)   Is the colored plane parallel to any coordinate planes? Which one(s)?

ii)   Does it intersect any of the axes? Which one(s)?

iii)   Write the coordinates of the intercepts.

iv)   What is the equation of the plane? Write your equation in the form: $ax + by = c$.

c)   Look at Model C as a group.

i)   Is the colored plane parallel to any coordinate planes? Which one(s)?

ii)   Does it intersect any of the axes? Which one(s)?

iii)   Write the coordinates of the intercepts.

iv)   Using what you have done previously, what do you think is the equation of the plane? How could you justify your thinking?

LS-69.   Make a table like the one below and fill in the coordinates of three points which are anywhere in the plane listed. Pick any points you like that are in the required location.

	xy-plane	xz-plane	yz-plane	points not in xy-, xz-, or yz-plane
1st point	(, ,)	(, ,)	(, ,)	(, ,)
2nd point	(, ,)	(, ,)	(, ,)	(, ,)
3rd point	(, ,)	(, ,)	(, ,)	(, ,)

a)   What do you notice about the coordinates of the points in the xy-plane?

b)   Make a conjecture about the coordinates of points in any of the three planes.

LS-70.   Using the table from the previous problem, answer the questions below.

a)   What is the equation of the xy-plane? That is, what equation is satisfied by all the points in the xy-plane?

b)   What is the equation for the yz-plane?

c)   What would the graph of $y = 0$ look like on a 3-dimensional graph?

d)   Draw the graph of $y = 0$ on isometric dot paper.

LS-71.   Look at Model B.

a)   Using the needlepoint grid as if it were a graphing grid, approximate some values for the points where the colored plane meets the xy-plane. What is the intersection of the colored plane with the xy-plane?

b)   Write an equation of the line formed by the intersection in part (a).

LS-72.   Look at Model C and do the following:

a)   Write an equation for the intersection of the yarn plane with the xy-plane.

b)   Write an equation for the intersection of the yarn plane with the yz-plane.

c)   Write an equation for the intersection of the yarn plane with the xz-plane.

LS-73.   Using isometric dot paper, sketch each graph on a separate set of 3-dimensional axes by finding x, y, or z intercepts. Sketch the graphs on the same sheet of paper.

a) $3x + 5y = 15$ as a   b) $3x + 5z = 15$ as a   c) $5y + 5z = 15$ as a
line in the xy-plane   line in the xz-plane   line in the yz-plane

d)   We now want to graph $3x + 5y + 5z = 15$. When $z = 0$, what plane are we in? Replacing z with zero, what equation will we have in this plane?

e)   Using the idea from part d), we have the same line as in part a). Continue with letting just $y = 0$ and then just $x = 0$, and we will have the lines you formed in b) and c). Draw all the lines on one set of axes. You now have a triangular piece of the plane represented by $3x + 5y + 5z = 15$. Shade this piece.

LS-74.   Graph x = 4:

a)   on a number line.

b)   in the xy-plane.

c)   in 3-space.

d)   How do the graphs differ in (a), (b) and (c)?

e)   Can you find the graph of (a) in the graph of (b)? Describe it.

f)   Can you find the graph of (b) in the graph of (c)? Describe it.

LS-75.   Three red rods are 2 cm longer than two blue rods. Three blue rods are 2 cm longer than four red rods. How long is each rod?

LS-76.   Sketch the graph of the following on separate pairs of axes

a) $y + 5 = (x - 2)^2$ b) $y \le (x + 3)^3$ c) $y = 4 + \dfrac{1}{x - 3}$

LS-77.   Rutilio built a pantry cabinet which was 24" x 32" x 93". He and his helper carried the cabinet on its side going through the customer's door. If the ceiling was 8' high will he be able to stand up the cabinet? Include a drawing and explanation with calculations to support your conclusion.

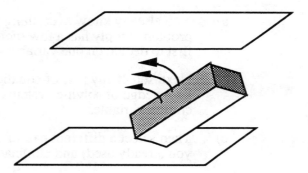

LS-78.   Solve the system:

$$x + 3y = 16$$
$$x - 2y = 31$$

Now, Rewrite the system and replace x with x^2. What effect will this have on the solution to the system?   Solve the new system.

LS-79.   The cost of food has been increasing about 4% per year for many years. To find the cost of an item 15 years ago, Jonique said "Take the current price and divide by $(1.04)^{15}$." Her friend Elissa said "That might work, but it's easier to take the current price and multiply by $(.96)^{15}$!" Explain who is correct and why.

LS-80.   Solve for w.
a) $w^2 + 4w = 0$ b) $5w^2 - 2w = 0$ c) $w^2 = 6w$

LS-81.   In 1976, the world's largest ice cream sundae was made in New York City's Central Park. It used 1500 gallons of ice cream and weighed 7, 250 lbs, 25 of which were cherries. One average sized cherry weighs approximately 6.5 grams and one kilogram is approximately equal to 2.2 pounds.

    a)   Approximately how many cherries were used for the sundae?

    b)   If the sundae was divided up evenly so that each person received one cherry, what would be the weight of an individual serving of the sundae?

SYSTEMS WITH THREE EQUATIONS & THREE VARIABLES

LS-82.   Suppose each equation in a system of equations such as those in the next problem represents a plane. Then the graph of the system will be three planes in space. Imagine three planes. In how many different ways could they intersect or not intersect? Describe the ways.

LS-83.   Consider the three equations, with three variables, shown below.

$$x + y - 2z = 5$$
$$2x + y + z = 0$$
$$3x - 2y + z = 1$$

    a)   You already know everything you need to know to solve this system. This problem simply has a few more subproblems. As with two equations you need first to decide on one variable to ELIMINATE.

    b)   Next, select any pair of equations from the above set of three. Use your knowledge of solving systems of equations with two variables to eliminate your chosen variable.

    c)   Then select a different pair of equations (the one you have not used yet and one you already used) and eliminate the same variable again.

    d)   You have modified the original system so that now you have a familiar situation; two equations with two variables. Rewrite and solve the new system.

    e)   You now know two of the three variables, but you are not finished. What is missing? Complete the solution.

    f)   Check your solution in each of the original equations.

    g)   What does your solution tell you about the graphs of the three planes.

LS-84.   What subproblems did you solve in the previous problem ?

LS-85.   Solve this system of equations and then check your solution in each equation. Be sure to keep your subproblems well organized.

$$x - 2y + 3z = 10$$
$$2x + y + z = 10$$
$$x + y + 2z = 14$$

LS-86. Write a system of inequalities for
        the graph at right.

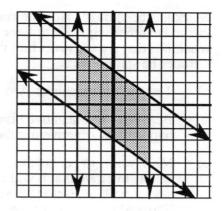

LS-87. Find reasonable equations which will generate each graph. You may want to look at
        your Parent Graph Tool Kit.

a)   b)   c)

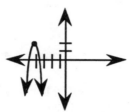   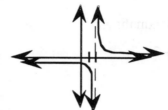   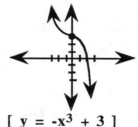

[y= -2(x+4)² +2]   [y = -x³ + 3]

                        [y = 1/(x - 2)

LS-88. Is $y = \dfrac{1}{x}$ the parent of $y = \dfrac{1}{x^2 + 7}$? Why?

LS-89. Solve each of the following for x:

        a)   2x + x = b   b)   2ax + 3ax = b   c)   x + ax = b

LS-90. What would happen graphically if a system of three equations had no solution?

LS-91.   Mark has been busy. Now he claims to have
          created another sequence of three function
          machines that always gives him the same
          number he started with.

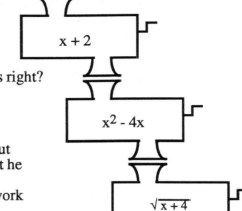

a)   Test out his machines. Do you think he is right?

b)   Be sure to test negative numbers. What
      happens for negative numbers?

c)   Mark wants to get his machine patented but
      he has to prove that it will always do what he
      says it will, at least for positive numbers.
      Show Mark how to prove his machines work
      by dropping in a variable and writing out
      each step the machines must take.

d)   Why do the negative numbers come out positive?

LS-92.   For this system: of inequalities:

$$y \le -2x + 3$$
$$y \ge x$$
$$x \ge -1$$

a)   Draw the graph.

b)   Find the area of the shaded region.

LS-93.   Given $f(x) = 2x^2 - 4$ and $g(x) = 5x + 3$ find:

a) $f(a) =$ b) $f(3a) =$ c) $f(a + b) =$

d) $f(x + 7) =$ e) $f(5x + 3) =$ f) $g(f(x)) =$

LS-94.   If a movie ticket now averages $6.75 and has been increasing an average of 15% per
          year, compute the cost:

a)   in 8 years   b)   8 years ago

LS-95.   Graph the solution of:

$$y > (x - 2)^2$$
$$y \le (x - 2)^3 + 2$$

LS-96.   You have a bag with 60 black jelly beans and 240 red ones.

a)   If you draw one jelly bean out of the bag, find the probability that it is black.

b)   If you add 60 black jelly beans to the original bag and draw out a bean, what is the probability that it is black?

c)   How many black beans do you need to add to the original bag to double the original probability of drawing a black bean?

d)   Write an equation that represents the problem in part (c).

THE EQUATION OF A PARABOLA FROM THREE POINTS

LS-97.   We already know how to find the equation of a line given two points. First we find the slope of the line and then we find the y-intercept. There is another way to think about this problem. If we consider the equation of any line as $y = ax + b$, we can solve this problem using a system of equations.

a)   The fact that a line goes through two points (-3, 4) and (-2, -1) means those values for (x, y) make the equation $y = ax + b$ true. Use this information to write two equations in which a and b are the variables.

b)   Use your equations from part (a) to find the equation of the line through the points (-3, 4) and (-2, -1). (That is, solve for a and b, not x and y.)

LS-98.   A parabola with a vertical line of symmetry can be represented by $y = ax^2 + bx + c$. It takes just two points to determine a line. How many points do you think it would take to determine such a parabola? Discuss the question with your group before doing the next problem.

a)   Use the idea you used in the previous problem about the line to find an equation for a parabola that passes through the points (2, 3), (-1, 6), and (0, 3). You can start by writing three equations in which a, b, and c are the variables.

b)   Solve the system of three equations and use the results to write the equation of a parabola.

LS-99.   Find the equation of the parabola through the points given:

a)   (3, 10), (5, 36), and (-2, 15).

b)   (2, 2), (-4, 5), and (6, 0).

LS-100.   What happened in part (b) in the last problem? Why did this occur? (If you are not sure, plot the points.)

LS-101.   Would it be possible to write an equation for a parabola for any three non-collinear points? Explain your thinking.

LS-102.   Write the system of inequalities
which will give you the graph at
right.   [y ≤ -x + 4; y > $^1/_3$x]

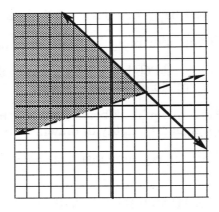

LS-103.   Arthur was supposed to solve the following system of equations.

$$5x + y + 2z = 6$$
$$3x - 6y - 9z = -48$$
$$x - 2y + z = 12$$

Arthur decided to solve this system by eliminating x. He started with the second and
third equations and realized that he could divide both sides of the second by -3 to get

$$-x + 2y + 3z = 16$$

He then combined this result with the third equation to get

$$
\begin{aligned}
-x + 2y + 3z &= 16 \\
\underline{x - 2y + z} &= \underline{12} \\
4z &= 28 \\
z &= 7
\end{aligned}
$$

"Wow! Two with one shot!" said Arthur. But then he didn't know what to do next.
What should he do to find x and y? Do it.

LS-104.   Solve for x: $1 - \dfrac{b}{x} = a$

LS-105.   Compare and contrast the graphs of $y = (x - 3)^2$ and $y = x^2 - 3$.

LS-106.   A stick two feet long is accidentally broken. [A diagram is essential making sense of
this problem]

a)   What is the probability that each of the pieces is at least 9 inches long?

b)   What is the probability that each of the pieces is at least x inches long?

LS-107.   Solve for integers x, y, and z: $\left(2^x\right)\left(3^y\right)\left(5^z\right) = \left(2^3\right)\left(3^{x-2}\right)\left(5^{2x-3y}\right)$.

LS-108.   Given $f(x) = 2x^2 - 4$ and $g(x) = 5x + 3$, find:

   a) $g(-2)$.

   b) $f(-7)$.

   c)   Evaluate $f\big(g(-2)\big)$ by first substituting for $g(-2)$.

   d)   How do you think you would calculate $f\big(g(1)\big)$? Find the value.

LS-109.   Find the measure $\angle CPM$.

   List any subproblems that were
   necessary to solve this problem.

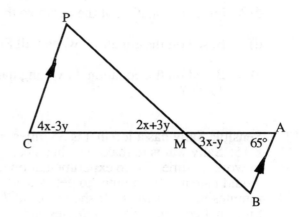

LS-110.   Describe how the graph of $y + 3 = -2(x + 1)^2$ is different from $y = x^2$.

WRITING EQUATIONS OF PARABOLAS FROM APPLICATIONS

LS-111.   A spaceship is approaching a star and is caught in its gravitational pull. When the
   ship's engines are fired the ship will slow down, momentarily stop, and then,
   hopefully, pick up speed, move away from the star, and not be pulled in by the
   gravitational field. The engines were engaged when the ship was 750 thousand miles
   away from the star. After one minute the ship was 635 thousand miles away. After
   two minutes the ship was 530 thousand miles away from the sun.

   a)   Find three points from the information above where x = the time since the engines
        were engaged and y = the distance (in thousands of miles) from the star.

   b)   Plot the points in part (a). We want the distance to reach a minimum and then
        increase again, over time. What kind of model follows this pattern?

   c)   Find the equation of the parabola that fits the three points you found in part (a).

   d)   If the ship comes within 50 thousand miles of the star, the shields will fail and the
        ship will burn up. Use your equation to determine whether the space ship has
        failed to escape the gravity of the star.

LS-112.   Sid Sickly has contracted an infection and has gone to the doctor for help. The doctor
           takes a blood sample and finds 900 bacteria per cc. ("cc" stands for cubic centimeter
           which is equal to 1 milliliter.) Sid gets a shot of a strong antibiotic from the doctor.
           The bacteria will continue to grow for a period of time, reach a peak, and then decrease
           as the medication succeeds in overcoming the infection. After 10 days, the infection
           has grown to 1600 bacteria per cc. After 15 days it has grown to 1875.

    a)   What are the three data points?

    b)   Make a rough sketch that will show the number of bacteria per cc over time.

    c)   Find the equation of the parabola that contains the three data points.

    d)   Based on the equation, when will Sid be cured?

    e)   Based on the equation, how long had Sid been infected before he went to the
           doctor?

LS-113.   Sensible Sally has a job that is 35 miles from her home. She needs to be at work by
           8:15. Sally wants to maximize her sleep time by leaving as late as possible but still get
           to work on time. From experimentation, Sally discovered that if she left at 7:10, it
           would take her 40 minutes to get to work. If she leaves at 7:30, it will take her 60
           minutes to get to work. If she leaves at 7:40, it will take her 50 minutes to get to work.
           Since her commute time increases and then decreases, Sally decides to use a parabola to
           model her commute. Assume the time it takes to get to work varies quadratically with
           the number of minutes after 7:00 that Sally leaves.

    a)   Let x = the number of minutes after 7:00 that Sally leaves.
           Let y = the number of minutes it takes Sally to get to work.

    b)   One ordered pair is (10, 40). What are the other two ordered pairs given in the
           problem?

    c)   Write three equations in three variables based on the general formula of the
           quadratic equation $ax^2 + bx + c = y$.

    d)   Solve the system to find a, b, and c. Write the particular quadratic equation for
           this function. Round your answer to 3 decimal places.

    e)   Use your equation to find how long it would take to get to work if Sally left at
           7:20

    f)   According to your equation, how long would it take to get to work if she left at
           7:58? Does this make sense? Why?

    g)   Find reasonable limitations for the domain of the function assuming the maximum
           average speed of 60 mph.

    h)   What time(s) does Sally need to leave to spend at most 45 minutes in her car?

    i)   What is the latest time, keeping in mind the restrictions on the domain, that Sally
           can leave in order to get to work by 8:15?

LS-114.   **PORTFOLIO: GROWTH OVER TIME - PROBLEM #2**

On a separate piece of paper (so you can hand it in separately or add it to your portfolio) explain **everything** that you now know about:
$$f(x) = 2^x - 3.$$

LS-115.   Write an explanation for someone just coming into the class about how to use subproblems to solve a system of four linear equations with four unknowns. Give such a clear explanation that your teacher becomes so convinced that you know what to do that he or she will never ask you to actually solve one.

LS-116.   **SUMMARY ASSIGNMENT**

a)   In this chapter you have looked at solving linear equations (and at when they don't have a solution) for systems with both two and three variables. Write a paragraph to explain the similarities and the differences of the <u>geometry</u> of the graphs for the two variable versus the three variable situations.

b)   On graph paper draw a polygon. Label its vertices then write the set of inequalities for which the intersection will be the area inside the polygon you drew. You can do an easy one, a triangle, or you can consider this a challenge and see how interesting a figure you can represent.

c)   What were the most difficult parts of this chapter? List sample problems and discuss the hard parts.

d)   What problem did you like best and what did you like about it?

LS-117.   Consider the pattern below:

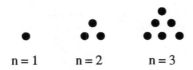

n = 1   n = 2   n = 3

a)   Draw n = 4?

b)   How many dots are in each figure from 1 to 4? Is this sequence arithmetic? Is this sequence geometric?

c)   Find the equation where, given n, you can determine the number of dots in the figure n. If you get stuck, try graphing the points (n, d) where d is the number of dots.

LS-118.   Graph each system and shade the solution:

a) $y \geq x^2 - 4$

      $y < \frac{1}{3}x + 1$

b) $y < 2x + 5$

      $y \geq |x + 1|$

LS-119.   Solve the system:

$$5x - 4y - 6z = -19$$
$$-2x + 2y + z = 5$$
$$3x - 6y - 5z = -16$$

LS-120.   a)   Find the equation of the parabola that passes through (2, 3), (-1, 6), and (0, 3)

b)   Find the vertex of the parabola.

LS-121.   **PORTFOLIO ASSIGNMENT**

a)   Find what you believe to be your two best pieces of work for this chapter.
Explain why you are particularly proud of these assignments. Be sure to include
a restatement of the original problem or a copy of it.

b)   Select two problems from the year that you are still having difficulty with. Copy
each problem and solve as much of it as you can. Explain what part of the
problem you don't understand.

FOLDING A 3-DIMENSIONAL AXIS SYSTEM

Origami Space: You will construct a three-dimensional axis system from folded paper. You need two squares each of two different colorsand two squares of graph paper. For each square:

a)   Fold the paper in half forming two congruent rectangles. Unfold, do not turn over. Keep the crease down. (This forms a valley fold.)

b)   Turn your paper 90° and repeat step (a), forming a total of 4 small squares.

c)   Flip your paper over so that all of the folds point upward.

d)   Fold the paper along the diagonal. The diagonals should be a valley while the others point upward (mountains). Unfold but keep the creases. Fold along the other diagonal. Unfold. You should now have eight triangles.

e)   Now push down in the center where the folds intersect. This should cause the diagonal folds to pop down and the other folds to pop up.

f)   Pinch together the two triangles at each of two opposite vertices. Push these vertices towards each other forming a "flower." You may want to re-crease your diagonal folds.

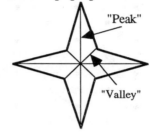

Assembly:

a)   Start with two flowers of different colors. Cover the "petal" of one flower with the petal of another. Pull the flowers apart slightly. Paper clip the covering petal.

b)   Take a graph paper flower and weave two of its adjacent petals with the other two flowers. Make sure that no <u>color</u> is a <u>cover</u> for more than one petal at this time. Add two more paper clips. You should now see an octant.

c)   Take any of the remaining flowers. Weave it with your structure so that each time you form an octant it has each of the three colors showing. Remove the paper clips and ease the flowers together. If they don't stay together, use a few paper clips.

d)   If you imagine each plane infinitely extended, into how many regions have we divided space?

e)   What is each region called?

f)   Hold the structure so you are looking at the graph paper plane. Now label this the x,y-plane by writing in the designations +x, -x, +y, -y.

Consider this axis set

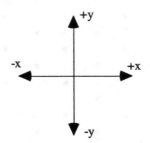

Looking at the graph
paper plane

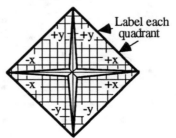

Isometric Grid

Chapter 6
Logarithms and Other Inverses
(The Case of the Cooling Corpse)

PROBLEM SOLVING

REPRESENTATION/MODELING

FUNCTIONS/GRAPHING

INTERSECTIONS/SYSTEMS

ALGORITHMS

REASONING/COMMUNICATION

CHAPTER 6
THE CASE OF THE COOLING CORPSE: LOGARITHMS AND OTHER INVERSES

In this chapter you will have the opportunity to:

- Undo everything you have done so far. Well not exactly, you will learn about inverse functions which reverse the order in which their original functions operate.

- Learn about a new function, the undoing or inverse function for an exponential function.

- Appreciate the usefulness of logarithms for solving some equations and other problems.

Again a graphing calculator will be a valuable tool and you will continue to need at least a scientific calculator for homework.

"UNDOING MACHINES"
Think about how you would solve this equation: $1.04^x = 2$

This might seem like a simple problem, after all, you saw similar equations back in Chapter 3. Try solving it now. Is it easy? Did you use guess and check? One of the goals of this chapter is to provide you with more of the algebraic tools you will need to solve bigger problems, including problems similar to these. Along the way, we hope you will increase your level of understanding of what a function is.

CC-1.   To the right is a picture of Anita's function machine. When she put a 3 into the machine, the machine put out a 7. When she put in a 4, the machine gave her a 9, and when she put in a -3, out came -5.

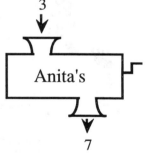

a)   Explain in words what this machine does to a number.

b)   Suppose it is possible to turn the crank the opposite direction, causing the machine to pull 7 back up into it. What do you think will come out the top? Explain.

c)   Suppose Anita wants to build another machine that will **undo** the effects of her first function machine. That is, she wants a machine that will take seven and turn it back into three, take nine and turn it back into four, and so on. Write a rule that she should program into this new machine to make it do this.

CC-2.   The function machine to the right **f** follows the rule:

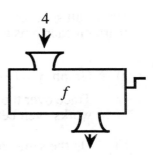

$$f(x) = 5x + 2.$$

a)   What is f(4)?

b)   If the crank is turned backwards, what number should be pulled up into the machine in order to have a 4 come out the top?

c)   Keiko wants to build a new machine that will undo what f does. What must Keiko's machine do to 17 to undo it and make it a 3? Write your rule in function notation and call it g(x).

CC-3.   Find the undo rules for each of the functions below. Write your answers in function notation, but be sure to use a name that is different from the function name. For example, if the function is named f(x), you can't use f(x) for the undoing machine too. You'll need to use some other letter, like g(x).

a) $f(x) = 3x - 2$

b) $h(x) = \dfrac{x + 1}{5}$

c) $p(x) = 2(x + 3)$

d) $q(x) = \dfrac{x}{2} - 3$

CC-4.   Diane claims that $f(x) = \dfrac{3}{x}$ is its own undo rule. Is her conjecture correct? Show how you know.

♀CC-5.   The formal mathematical name for an undoing function is **inverse**. Record this in your Tool Kit! Find the inverse for each of the functions below. Then graph each function and its inverse on the same set of coordinate axes. In other words, use one set of axes for part (a), a new set for part (b), etc. This is NOT a good time to divide up the graphing among your group members. Make sure each of you does the work for each part. Check with each other as you go.

a) $f(x) = 2x + 4$

b) $f(x) = -\dfrac{2}{3}x$

c) $y = \dfrac{1}{3}x + 2$

d) $y = x^3 + 1$

CC-6.   When you have completed all the pairs of graphs, look for patterns in the graphs. What relationships do you see between the graph of a function and the graph of its inverse? Do the pairs of graphs have a line of symmetry? Justify your answer.

CC-7.   Use a full sheet of graph paper. Put the axes in the center of the page, and label each
            mark on each axis as one unit. Use pencil (the softer the lead the better.)

    a)   Graph $y = (\frac{x}{2})^2$ over the domain $0 \le x \le 8$. Label the graph with its equation.
            Trace over the graph with the pencil until the graph is heavy and dark. Crayon
            works even better than pencil.

    b)   On the same graph paper, graph $y = x$ using ball-point pen (use a straightedge,
            and press down hard).

    c)   What is the equation of the inverse of this parabola?

    d)   Using a ruler as an edge, fold the paper along the line $y = x$, with the graphs on
            the **inside** of the fold. Put the folded paper on your desk top. You will be able
            to see the darkened graph of the half-parabola through the paper. Now press the
            paper with a hard object, such as the back of your fingernail or the top of a pen,
            so that you are rubbing the graph from the back of the paper. The purpose is to
            make a "carbon copy" of the graph of the parabola, reflected across the line.
            Then open the paper and fill in the picture if it is not completely copied. Write
            your observations on the graph.

    e)   Find three points (x, y) that satisfy the equation you wrote in part (c). Find each point
            on the "carbon copy" graph. Each should be a point on this graph. Explain why.

CC-8.   Antonio's function machine is shown at right:

    a)   What is A(2)?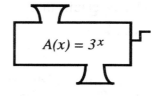

    b)   If 81 came out, what was dropped in?

    c)   If 8 came out, what was dropped in? Be accurate to two decimal places.

CC-9.   If $10^x = 10^y$, what is true about x and y? Justify your answer.

CC-10.   Sketch the solution of this system of inequalities.

            $$y > x^2 - 5$$

            $$y < -(x - 1)^2 + 7$$

CC-11.   Graph $y = \frac{1}{2}x - 3$ and its undoing function on the same set of axes.

    a)   What is the equation of the undoing function?

    b)   Does this graph, with both lines on it, have a line of symmetry? If so, what?

CC-12.   Solve for x: $x^2 = 91$.

CC-13.   Solve the equation $3 = 8^x$ for x, accurate to two decimal places.

CC-14.   Solve the following system:

$$x + y + z = 2$$
$$2x - y + z = -1$$
$$3x - 2y + 5z = 16$$

CC-15.   Dana's mother gave her $175 on her sixteenth birthday. "But you must put it in the bank and leave it there until your eighteenth birthday," she told Dana. Dana already had $237.54 in her account, which pays 3.25% annual interest, compounded quarterly. Give a lower bound on how much money she will have on her eighteenth birthday if she makes <u>no</u> withdrawals before then. Justify your answer.

CC-16.   If $2^{x+4} = 2^{3x-1}$, what is x?

CC-17.   Write the equation of a circle with a center at (-3, 5) that is tangent to the y-axis. Sketching a picture will help.

CC-18.

REPORT ON GROWTH
A two-page project description is included at the end of this chapter. Read it now and come to class prepared with any questions or ideas.

INVERSE TO INTERCHANGE
CC-19.   Wanda and Samantha have found a useful pattern between functions and their inverses. They didn't remember any shortcuts for graphing lines, so they made a table to graph $f(x) = 2x + 7$ and its inverse. That's when they made their discovery.

a)   Make the two tables and see if you can find their pattern.

b)   If (10, 27) is a point on the graph of the function, what point do you automatically know to be on the inverse graph? What if (3.5, 14) is on the function's graph? What if (a,b) is on the graph? What about (x, y)?

c)   Discuss the significance of this discovery with your group before going on to the next problem.

CC-20.   Kalani began to think that this method of **graphing** the inverse of a function might
          help him find the **equation** of the inverse. "If you can just interchange x and y to find
          points on the graph of the inverse, why not just switch x and y in the equation itself to
          find the equation of the inverse?" he said to his friend Macario.

          "Well, I think I see what you are saying. A function and its inverse are reflected across
          the line y = x, and in the table of values, the values for the independent and dependent
          variables are just switched. But what makes you think we can do the same thing with
          the equation?" Macario asked.

          "Think about it: an equation represents **all** the points on the line, so if you can
          interchange all of the x and y values, why not interchange the x and y variables?"

          a)   Try this method of interchanging the x and y variables on y = 3x - 1. Solve the
                new equation for y.

          b)   Did this method work? Show how you know.

          c)   The inverse of a function could be called the **x-y interchange** of the function.
                Give a good reason why some one might want to call it that.

CC-21.   Find the inverse of each of the functions below. Write your answers in function
          notation. Remember to use a new name for the new function!

          a)   f(x) = 8x + 6   b)   f(x) = 3x^5

CC-22.   Uyregor thinks his instructor is sadistic!

          He gave Uyregor the function $f(x) = \frac{3}{5}x + 9$ and told him that he is supposed to find
          the inverse of this function and call it g(x). That's the easy part. Once he has done
          that, he is supposed to find f(1), f(2), f(3), . . . , f(50)! But wait! That's not all!
          Then he is supposed to plug those results into g(x)! Since you have such a **nice**
          teacher, you can help Uyregor out. Explain to him why his instructor is not so mean
          after all and how he should already know all of the final results.

CC-23.   The graphs of some functions are sketched below. For each function, sketch the graph
          of the inverse. Then state the domain and range for each function and for its inverse.
          The dotted line is y = x.

          a)   b)   c)

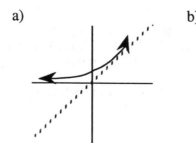   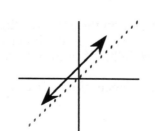   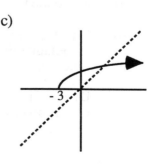

CC-24.   Trejo says that if you know the x and y intercepts, domain, and range of an equation, then you automatically know the x-intercepts, y-intercepts, domain, and range for the inverse. Hilary disagrees. She says you know the intercepts, but that is all you know for sure. Who is correct? Justify your answer.

CC-25.   Lacey and Richens each have their own personal function machines. Lacey's, L(x), squares the input and then subtracts one. Richens' function, R(x), adds 2 to the input, and then multiplies by three.

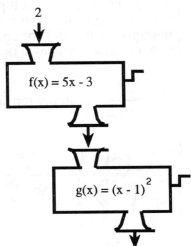

a)   Write the equations that represent L(x) and R(x).

b)   Lacey and Richens decide to connect their two machines, so that Lacey's output becomes Richens' input. Eventually, what is the output if 3 is the initial input?

c)   What if the order of the machines was changed. Would it change the output? Justify your answer.

CC-26.   Two function machines, $f(x) = 5x - 3$ and $g(x) = (x - 1)^2$, are shown at the right.

Suppose f(2), whatever that equals, is dropped into the g(x) machine. This is written as $g\big(f(2)\big)$.

What is this output?

CC-27.   Using the same function machines as in the previous problem, what is $f\big(g(2)\big)$? Careful! The result is different from the last one because the order in which you use the machines has been switched! With $f\big(g(2)\big)$, first you will find g(2), then you substitute that answer into the f machine.

CC-28.   When we push two (or more) function machines together, we say we have a new function which is the **composition** of the two functions. The composition can be written different ways, as $f\big(g(x)\big)$ or sometimes as $f \circ g(x)$. Does it seem to matter in what order you use the functions? Does $f\big(g(x)\big) = g\big(f(x)\big)$? Explain.

CC-29.   Rebecca thinks that she has found a quick way to graph an inverse of a function. She figures that if you can interchange x and y to find the inverse, she will interchange the x and y axes by flipping the paper over so that when she looks through the back the x-axis is vertical and the y-axis is horizontal as shown below: Copy the graph on the right on a separate sheet and try her technique. What do you think?

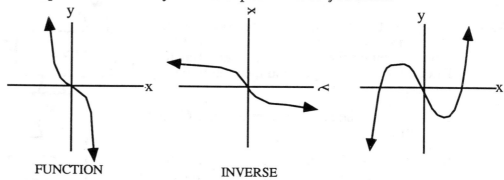

FUNCTION   INVERSE

♀CC-30.   FRACTION BUSTERS! You can use fraction busters to simplify complicated expressions like the one below. Instead of multiplying both sides by the common denominator, we can multiply the numerator and denominator of the large fraction by the common denominator of the smaller fractions. In the example below, we can multiply the top and bottom by **ab** to clear the fractions in the numerator and the denominator. (provided neither a nor b is equal to 0)

$$\frac{\frac{a}{b}}{1 - \frac{1}{a}} \quad \text{multiply top and bottom by } \mathbf{ab} \text{ (common denominator of all the fractions)}$$

$$\frac{\left(\frac{a}{b}\right) \cdot \frac{ab}{1}}{\left(1 - \frac{1}{a}\right) \frac{ab}{1}} = \frac{\frac{a^2 b}{b}}{ab - \frac{ab}{a}} \quad \text{Now simplify.}$$

$$\frac{\frac{a^2 \cancel{b}}{\cancel{b}}}{ab - \frac{\cancel{a}b}{\cancel{a}}} = \frac{a^2}{ab - b}$$

Use this technique to simplify each problem below:

a) $\dfrac{x}{1 - \dfrac{1}{x}}$

b) $\dfrac{\dfrac{1}{a} + \dfrac{1}{b}}{\dfrac{1}{b} - a}$

CC-31.   Sketch the graph of $(x - 2)^2 + y^2 = 20$

CC-32.   Samy has a 10 foot ladder which he needs to climb to reach the roof of his house. The roof is 12 feet above the ground. The base of the ladder must be at least 1.5 feet from the base of the house. If he feels safe standing on the top step of the ladder, how far will he have to step up to get onto the roof? Draw a sketch.

CC-33.   Later, after Samy had recovered from his fall off the roof, a fall which broke his ladder into two pieces, he needed his ladder to reach the top of a window he was washing. He needed a ladder that was at least four feet long. At the same time, his neighbor, Elisa, needed to borrow a ladder, and she **also** needed one that was at least four feet long. What is the probability that Sam's fall, which broke the ladder into two random pieces, caused the ladder to break in a spot so that both he and Elisa would be able to use a ladder at least four feet long?

CC-34.   Danielle was having trouble finding the undo function (the inverse) for $g(x) = x^2$. So she graphed the function $y = x^2$ and used the carbon copy method to get the interchanged graph at right.

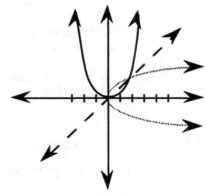

a)   "Aha", she said, "now I know the interchange function." What equation was she thinking of?

b)   "But wait!" said Regis. "$x = y^2$ is not a function." "Picky, picky, picky," Danielle replied, "it works doesn't it?" What do you think? Why can't $y = x^2$ have an inverse <u>function</u>?

c)   Think of another function that does not have an inverse function. Discuss this with your group.

CC-35.   As you saw in the previous problem some functions do not have inverse <u>functions</u>. There may be an inverse equation that works, but it is not a function. Sometimes when this happens we "cheat." In order to get an inverse <u>function</u> we use only a part of the original function. Problem CC-34 is an example.

For each of the following graph the original function and its inverse function and find the equation of the undoing function.

a) $f(x) = x^2$ with domain $x \geq 0$

b) $g(x) = (x - 2)^2$ with domain $x \geq 2$

CC-36.   Graph each of the following functions on a separate pair of axes. Clearly label the equation of the graph.

i) $y = \dfrac{-2}{3}x + 6$

ii) $y = \dfrac{1}{2}(x + 4)^2 + 1$ Domain $x \geq -4$

iii) $y = \dfrac{1}{x}$

a)   For each of the functions above, state the x-intercepts, y-intercepts, domain, and range.

b)   Graph the inverse of each function by using the "carbon copy" method or by interchanging points. Clearly label the inverse.

c)   State the x-intercepts, y-intercepts, domain, and range for each of the inverse graphs.

d)   Find the "undoing" equation for each and write it next to the graph of the inverse function.

♀LS-37.   Consider the graph of the equation $y = \sqrt{x}$.

a)   Change the equation so its graph is stretched and moved 6 units to the right.

b)   Change the equation so its graph is compressed and moved 9 units down.

c)   Write a general equation for the family of functions that have $y = \sqrt{x}$ as a parent.

d)   For the general equation that you found in (c), make one non-trivial specific example and draw the accurate graph. Include all important information.

e)   Add "Square Root" to your Parent Graph Tool Kit.

CC-38.   Amanda's favorite function is $f(x) = 1 + \sqrt{x + 5}$. She has built a function machine that performs these operations on the input values. Her brother Eric is always trying to mess up Amanda's stuff, so he created the inverse of f(x), called it e(x), and programmed it into a machine.

a)   What is the equation of e(x), Eric's function, which is the inverse of Amanda's? (Remember, it must <u>undo</u> Amanda's machine.)

b)   What happens if the two machines are pushed together? What is $e\big(f(-4)\big)$? Explain why this happens.

c)   If f(x) and e(x) are graphed on the same set of axes, what would be true about the two graphs?

d)   Draw the two graphs on the same set of axes. Be sure to notice the restricted domain and range of Amanda's function.

CC-39.   If $g(x) = 4x + 7$, find the equation of a machine $f(x)$ so that $f(x)$ is the inverse of $g(x)$. Explain completely how you got your function and show that it works on at least three different numbers.

CC-40.   Solve the system below:

$$a - b + 2c = 2$$
$$a + 2b - c = 1$$
$$2a + b + c = 4$$

What does the solution tell you about the graphs?

CC-41.   Sketch the solution to this system of inequalities.

$$y \geq (x + 5)^2 - 6$$
$$y \leq (x + 4)^2 - 1$$

CC-42.   Make a sketch of a graph showing the change in temperature of a cup of coffee as it sits on the kitchen table to cool.

a)   What are the independent and dependent variables?

b)   What is the domain and range?

c)   Is there an asymptote? Explain.

CC-43.   Solve each of the following for x.

a) $x^2 = 27$ b) $2^x = 27$

CC-44.   Brian keeps getting negative exponents and fractional exponents confused. Help him by explaining the difference between $2^{1/2}$ and 2^{-1}.

CC-45.   Eniki has a sequence of numbers given by the formula $t(n) = 4(5^n)$.

a)   What are the first three terms of Eniki's sequence?

b)   Chelita thinks the number 312,500 is a term in Eniki's sequence. Is she right? Justify your answer by either giving the term number or explaining why it isn't.

c)   Elisa thinks the number 94,500 is a term in Eniki's sequence. Is she right? Explain.

CC-46.   **SILENT BOARD GAME**

Do this individually and silently (don't discuss it with your group). Figure out the rule for this function. The table is written on the board. As you figure out a number that fits, go up to the board and fill in the number (silently). If it's not the correct number, the teacher will erase it. Only one number from each person, please. Copy the table at the side of a full sheet of graph paper.

x	8	$\frac{1}{2}$	32	1	16	4	3	64	2	0	0.25	-1	$\sqrt{2}$	0.2	$\frac{1}{8}$
g(x)	3	-1		0				6							

CC-47.   When the SILENT BOARD GAME table is filled in, discuss in your groups how you found the entries and write a brief description of the method.

a)   Write your rule for the function in symbols.

b)   x = 1024 is not in the table. What is g(1024)?

CC-48.   Make a table like the one below and fill in the missing numbers.

a)

x	f(x)
81	4
$\frac{1}{3}$	
3	
27	
1	0
-1	
0	
9	2
$\frac{1}{9}$	
$\sqrt{3}$	

b)   Consider the equation for the function in part (a). There are two ways to write this equation; one uses exponents, and one uses the word "log". Write the equation in both ways.

CC-49.   Using a full sheet of graph paper, put the axes in the center of the page and label each mark on each axis as one unit.

a)   Draw the graph of the function from the SILENT BOARD GAME by plotting the table of values. On the same axes, in a contrasting color, graph $y = 2^x$ and label it with its equation.

b)   What would happen if you reflected the graph $y = 2^x$ across the line y = x? Reflect it by the "carbon copy" method if you would like to check it. On the graph, answer this question and write any other observations you can make about the two graphs.

CC-50.   We will call the function, $x = 2^y$ the **inverse exponential function, base 2**. Label the graph with this name. Then describe this function as completely as possible, more precisely, investigate it. Write all the important information on the graph paper.

CC-51.   Suppose we label the inverse exponential function, base 2, g(x). Refer back to the Silent Board Game table and your graph to evaluate g(x) = y below. What is y and/or x in each of the following cases?

a) $g(32) = y$ b) $g\left(\frac{1}{2}\right) = y$ c) $g(4) = y$

d) $g(0) = y$ e) $g(x) = 3$ f) $g(x) = \frac{1}{2}$

g) $g\left(\frac{1}{16}\right) = y$ h) $g(x) = 0$ i) $g(8) = y$

CC-52.   Sketch the graph of $y + 3 = 2^x$.

a)   What is the domain and range of this function?

b)   Does this function have a line of symmetry? If so, what?

c)   What are the x and y intercepts?

d)   Change the equation so that the graph of the new equation has no x intercepts.

CC-53.   What is the equation of the inverse of $f(x) = \sqrt{5x + 10}$? Make a graph of both the function and its inverse on the same set of axes.

CC-54.   Write the equation of an increasing exponential function, which has a horizontal asymptote at $y = 15$.

CC-55.   On Jason's math homework last night, he had to solve for x in several equations. He did fine on all of the problems involving numbers, but he just didn't believe that methods he used for problems such as this:

$$5 \cdot 4 + 2(x)^3 = 8 + 7$$

would work on problems such as this:

$$ay + bx^3 = c + 7.$$

Show Jason how he can use the same basic steps he used for the first problem to solve the second.

CC-56.   What is the difference between the graphs of :

$$y + 3 = (x - 1)^2 \quad \text{and} \quad x + 3 = (y - 1)^2?$$

Explain completely enough so that someone who does not know how to graph either, could graph them after reading your description. Are the two graphs inverses of each other? Are they both functions? Justify your answer.

CC-57.   Consider the sequence given in the table below:

n	0	1	2	3	4
t(n)	4	2	2	4	8

a)   Plot the points and draw in staircases to help you decide what type of sequence it is.

b)   Write the equation that represents this sequence. If you have trouble getting started look back to LS-97 or 98.

CC-58.   Which is larger, n - 1 or n - 2 ? Is your answer the same regardless of the value you chose for n? Explain.

CC-59.   Solve each of the following for x.

a) $x^3 = 243$ b) $3^x = 243$

CC-60.   Write the equation for each circle graphed below:

a)   b)

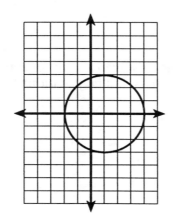   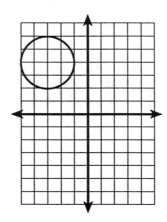

INTRODUCING LOGARITHMS!

CC-61.   If $x = 2^y$, how can you solve for y and make it say "y ="? Discuss this with your group and be prepared to report to the class what you think.

CC-62.   When mathematicians cannot solve a problem or do not have the tools to do a particular problem, do you know what they usually do? They invent a tool to do what they want! In this case they will invent a new operation! The inverse exponential function, base 2. Tadaa! Henceforth they said (at one point in history) the inverse exponential function base 2 will be known as **logarithm, base 2**. Why "logarithm" when thousands of other names were possible? That's an extra credit problem to check out with your teacher. You graphed this function back in problem CC-50 (we called the graph the inverse exponential function, base 2). Find your graph and label it as logarithm, base 2. Add the graph of $y = \log_2 x$ to your Parent Graph Tool Kit.

Logarithm base 2 is usually abbreviated as log, base 2 and is usually written $\log_2$. When we see this symbol, we read it aloud as "log, base 2." The comma means we pause a bit when we say it. Since it is a function, we can use function notation, as in $g(x) = \log_2(x)$. Notice that the base number, 2 in this case, is always written a little lower, as a subscript. Use your graph of g(x) to find each of the missing values:

a) $\log_2(32) = ?$

b) $\log_2\left(\frac{1}{2}\right) = ?$

c) $\log_2(4) = ?$

d) $\log_2(0) = ?$

e) $\log_2(?) = 3$

f) $\log_2(?) = 1/2$

g) $\log_2\left(\frac{1}{16}\right) = ?$

h) $\log_2(?) = 0$

i) $\log_2(8) = ?$

CC-63.   How do your answers in the previous problem relate to your answers in CC-51?

CC-64.   Let $y = \log_2(x)$. Rewrite this so that it is solved for x. Think about how we defined $y = \log_2(x)$ for a hint. Put a large box around both equations. Do the two equations look the same? Do the two equations mean the same thing? Are they equivalent? How do you know? This is very important. Don't rush, think about it.

CC-65.   Since a logarithm is the interchange (or inverse) of an exponential function, each logarithmic function has a particular base. Note that we write the base <u>below</u> the line (like a subscript). For example, we write $\log_2(x)$. This looks a bit confusing, since the "x" is above the 2. It almost looks like 2^x , but it isn't. When you write it, make it very clear. Every log equation can be written as an exponential equation and vice versa, as you saw in the previous problem. Copy each equation shown below. Then rewrite each equation in the other form.

a) $y = 5^x$

b) $y = \log_7(x)$

c) $8^x = y$

d) $A^K = C$

e) $K = \log_A(C)$

f) $\log_{1/2}(K) = N$

CC-66.   On Wednesdays at Tara's Taqueria four tacos are the same price as three burritos. Last Wednesday the Lunch Bunch ordered five tacos and six burritos, and their total bill was $8.58 (with no tax or drinks included). Nobody in the Lunch Bunch can remember the cost of one of Tara's tacos. Help them figure it out.

CC-67.   Delores has two big bags of hard candies that she is planning to put into small packages for the Ides of March picnic. One bag contains 450 assorted fruit flavored candies; the other contains 500 coffee-toffees. She figures that the children will like the fruit candies better, and the adults will prefer the coffee-toffees. So she plans to make little bags of 2 coffee-toffees and 6 fruit candies for the children and of 5 coffee-toffees and 3 fruits for the adults. Now she is trying to figure out what numbers of child and adult candy packets she can make. Help her out by writing two inequalities and drawing their graphs. Let x be the number of bags for children and y the number for adults. If sixty children and 70 adults go to the picnic will there be enough bags of candy?

CC-68.   Shakeah's grandfather is always complaining that back when he was a kid, he used to be able to buy his girlfriend dinner for only $1.50.

   a)   Assuming he wasn't just a cheap date, why would this be true?

   b)   If that same dinner, that Shakeah's grandfather purchased for $1.50 sixty years ago, now costs $25.25, and the increasing amounts form a geometric sequence, write an equation which will give you the costs at different times.

CC-69.   Make a sketch of a graph that is a decreasing exponential function with the x-axis as the horizontal asymptote.

   Now make a similar sketch, but this time the horizontal asymptote is the line $y = 5$.

CC-70.   If $f(x) = x^4$ and $g(x) = 3(x + 2)$, what is:

   a) $f(2)$?   b) $g(2)$?

   c) $f(g(2))$?   d) $g(f(2))$?

   e)   If $f(x) = 81$, what is x?

CC-71.   Suppose $h(x) = 3^x$ and $k(x) = \log_3(x)$. What is $h(k(x))$? What about $k(h(x))$? Start with a few numbers to convince yourself. Explain completely why this is true.

CC-72.   At the **Write-a-Text Factory,** workers spend grueling hours at computer terminals trying to be creative writing text books. One day, during a brief bout of boredom, Karna and Carlos decided to play a trick on Darell as he worked on his document. They implanted a strange code into his computer so that as soon as he had 60,000 characters 10% of his document, starting at the beginning, would be deleted every hour. Just as the clock struck 5:00 and he was anxiously waiting for the whistle to blow telling him he could go home, Darell typed in his 60,000th character! He left, not to return till 8:00 a.m. the next day.

   a)   How many characters were left when he returned the next morning?

   b)   If Darell starts typing 3600 characters per hour when he arrives, will he increase the size of his document faster than the mutant code can decrease it, or vice-versa?

CC-73.   A triangle is formed by a line which has a slope of 2, and the x and y axes. The area of the triangle is 30 square units. Find the equation of the line. A diagram would be helpful.

CC-74.   Eeew! Ever eat a maggot? Guess again! The FDA publishes a list, the Food Defect Action Levels list, which indicates limits for "natural or unavoidable" substances in processed food (*Time*, October 1990). So in 100 g of mushrooms, for instance, the government allows 20 maggots! The average rich and chunky spaghetti sauce has 350 grams of mushrooms. How many maggots is that?

CC-75.   If $x = 3^y$, what do you think you would write to solve for y? Explain.

INVESTIGATING LOGARITHMS

CC-76.   There is a "log" key on your calculator. On a normal scientific calculator, you enter the number first, then press the log key. On the graphing calculator, however, it is entered just as you would write it or say it (press "log" first, then the number). But you will notice that the base is omitted. Unfortunately, you do not get the option of putting in any base you want. Instead, the calculator's log has a fixed base. Your mission is to figure out this base. As a hint, try entering log(2), log(3), etc.. Keep track of your results. Use complete sentences to write an explanation of the problem and your solution.

CC-77.   Investigate $y = \log_b(x)$. Be complete and be sure to include several appropriate graphs.

♡CC-78.   When we compute with logarithms using the calculator, what base must we use?

> Since the calculator's base for logs is always 10, and since the calculator is so useful in computations, we save time (and possibly confuse students) by making this agreement: When we write log(x), *without writing a base*, we mean $\log_{10}(x)$.
>
> Log is a function, and like any function its inputs should be in parentheses like f(x). But since logs are so useful and written so frequently, most people omit the parentheses, and just write log x (or $\log_N$ x, if the base is not 10).

CC-79.   Copy these and solve for **x**. Obtain a numerical answer without a calculator if possible.

a) $\log_5(25) = x$

b) $\log_4\left(\frac{1}{4}\right) = x$

c) $3 = \log_x(343)$

d) $\log_6(0) = x$

e) $3 = \log_5(x)$

f) $\log_9(x) = \frac{1}{2}$

g) $x = \log_{64}(8)$

h) $\log_{11}(x) = 0$

i) $x = \log_{10}(0.01)$

♡CC-80.   Your friend is not quite sure how to rewrite exponential equations as log equations, and vice-versa. Write out an explanation for your friend. It might be easier to use equations, symbols, and letters of the alphabet in addition to sentences. What you are really doing here is developing a definition of logarithm. If you need help, rewrite the equations from the previous problem without the "log" to see a pattern.

CC-81.   Late last night, as Agent 008 negotiated his way back to headquarters after a long day, he saw the strangest glowing light. It came closer and closer until finally he could see what was creating the light. It was some kind of spaceship! He was frozen in his tracks. It landed only 15 feet from him, and a hatch slowly opened. Four little creatures came out carrying all sorts of equipment, from calculators to what appeared to be laser beams. They did not seem to notice that Agent 008 was there; that is, until he sneezed! Suddenly, the creatures turned around, looking very startled. They dashed into the spaceship, closed the hatch, and rocketed into the night. Could he believe what he had seen? Maybe it was just a dream? After a few minutes of standing there dazed and confused, he started walking on, his eyes glazed over. He was just coming to his senses when he stepped on something strange. He picked it up and to his surprise it was one of the creature's calculators! What a prize! He started playing with it as he walked on. Boy! HQ is going to love this! It appeared to have a log button and as he played with it he noticed something interesting: log 10 did not equal 1 as it did on his calculator. With this calculator, log 10 ≈ 0.926628408! He tried some more: log 100 ≈ 1.853256816, and log 1000 ≈ 2.779885224. This was most peculiar. Obviously, the creatures did not work in base 10!

    a)   What base do the space creatures work in? Explain how you got your answer. You may want to rewrite the problem as $\log_b 10 = 0.926628408$ and try to figure out the value of b.

    b)   How many fingers do these space creatures have?

CC-82.   Solve for x. Your answer must be accurate to three decimal places.

$$2 = 1.04^x$$

CC-83.   If $10^{3x} = 10^{x-8}$, solve for x. Show that your solution works by checking your answer.

CC-84.   For $f(x) = 3 + \sqrt{2x - 1}$, do each of the following:

    a)   What are the domain and range for f(x)?

    b)   What is f(x)'s inverse, or interchange? Call it g(x).

    c)   What are the domain and range of g(x)?

    d)   Calculate $f(g(6))$.

    e)   Calculate $g(f(6))$. What do you notice? Why does this happen?

CC-85.   Which of the following are true? Show why it is false if it is not true.

a) $\dfrac{x+3}{5} = \dfrac{x}{5} + \dfrac{3}{5}$

b) $\dfrac{5}{x+3} = \dfrac{5}{x} + \dfrac{5}{3}$

CC-86   Solve:

a) $5 - \dfrac{8}{x+4} = \dfrac{2x}{x+4}$ b) $\dfrac{1}{x} = \dfrac{x}{1-x}$

CC-87.   While working on his math homework, Pietro came across this problem:

"$F(x) = 3^{(-x)} - 1$, find $F(2)$, $F(3)$, $F(4)$ and $F(5)$. Then explain what happens as bigger and bigger numbers are substituted in for x."

He didn't have any idea how to do the problem.

a)   Help him out by calculating $F(2)$, $F(3)$, $F(4)$ and $F(5)$.

b)   Plot the points and draw the graph of $F(x)$.

c)   Use your graph to explain to Pietro what does happen to $F(x)$ as x gets larger and larger.

CC-88.   Solve for m: $m^5 = 50$.

CC-89.   Simplify the following:

a) $\dfrac{\frac{1}{a} - \frac{1}{b}}{\frac{1}{a} + \frac{1}{b}}$ b) $\dfrac{x+y}{\frac{1}{x} + \frac{1}{y}}$

CC-90.   A dart hits each of these dart boards at random. What is the probability that the dart will not land in the shaded area?

a)   b)

GRAPHS OF LOGARITHMIC FUNCTIONS

CC-91.   Using a full sheet of graph paper, make a fairly accurate graph of $f(x) = \log(x)$. Clearly write the equation on the graph.

♀CC-92.   On the same set of axes, but in another color, graph $g(x) = 5 + \log(x)$. How is this graph different from the first? Explain. You may use the graphing calculator, but if you use your Tool Kit, referring to notes from Chapter 4, you can probably do these graphs more efficiently without it.

CC-93.   On the same set of axes, but in yet another color, graph $h(x) = 5\log(x)$. How is it different? Explain.

CC-94.   You know the routine, graph $j(x) = \log(x + 5)$. How is it different? Explain.

CC-95.   Last one! Graph $k(x) = \log(x - 5)$. How is it different? Explain.

CC-96.   Prove that you are the expert log grapher! Explain completely, so that even a friend who hates math can follow, how to graph $y = 2 + 7\log(x + 4)$.

CC-97.   Graph $y = |\log(x)|$.

CC-98.   Copy these, and solve for x without a calculator:

a) $\log_x(25) = 1$ b) $x = \log_3(9)$ c) $3 = \log_7(x)$

d) $\log_3(x) = 1/2$ e) $3 = \log_x(27)$ f) $\log_{10}(10000) = x$

CC-99.   Calculator Investigation:

a)   Compute the decimal values for the following pairs of logarithmic expressions.

   $4 \log (2)$ and $\log (2^4)$ $2 \log (3)$ and $\log (9)$ $3 \log (4)$ and $\log (4)^3$

b)   What do you notice in all three examples? Explain completely. Rewriting the number within the parentheses as a power might help.

c)   Try $\log 1000^5$ and $5 \log 1000$ and $\log (10^3)^5$.

d)   Make up three examples of your own; use unusual numbers in one of them.

e)   Make a note of this pattern. You will use it again soon.

CC-100.   Solve for x. Your solution must be correct to **four** decimal places.

$$2 = 1.04^x$$

CC-101   Solve these systems of equations. Look for a short cut based on good algebraic reasoning. If you use a shortcut be sure to explain your reasoning.

    a) $13x + 17y = 0$ b) $435x + 334y = 0$
          $8x - 10y = 0$ $1.78x - 37.7y = 0$

CC-102.   Is it true that $\log_3(2) = \log_2(3)$? Justify your answer.

CC-103.   Consider the general form of an exponential function: $y = km^x$.

    a)   Solve for **k**.

    b)   Solve for **m**.

CC-104.   Graph the following two functions on the same set of axes.

$$y = 3(2^x)$$

$$y = 3(2^x) + 10$$

    a)   How do the two graphs compare?

    b)   Suppose the first equation is $y = k \cdot m^x$ and the graph is shifted up **b** units, what is the new equation?

CC-105.   Write the equations of two different curves that cross the line $y = 10$ but do not cross the line $y = 11$. Make sketches of each.

CC-106.   Given the function $y = 3(x + 2)^2 - 7$, how could you restrict the domain to give "half" of the graph?

    a)   Find the equation for the inverse function for your "half - a - function."

    b)   What are the domain and range for the inverse function?

USING LOGARITHMS TO SOLVE PROBLEMS

CC-107.   Discuss with your group how you solved $2 = 1.04^x$ accurately to four decimal places. Lay out the subproblems or key ideas.

CC-108.   Aren't you getting tired of all this guess and check? Don't you **wish** there were an easier way of doing this sort of problem? Well, you guessed it, there is, and actually you know most of the pieces already. You just need to put some ideas together! Converting this equation from exponential form to log form gives x = $\log_{1.04}(2)$, but our calculators are still useless, so this approach leads to a dead end.

Look back to CC-99 and make sure you see the pattern. This will help you in rewriting each of the following and is a step toward solving our "favorite problem." The important thing to notice from those problems is that if, for example, $\log 4^3 \approx 1.806$ and $3\log 4 \approx 1.806$, what can you conclude? Use this idea to write each of the following in a different form.

a) $\log 2^5 = ?$

b) $\log 10^5 = ?$

c) $\log 10^x = ?$

d) $\log 3^x = ?$

♀CC-109.   Does the pattern of CC-99 always hold? What follows is a demonstration that the pattern holds regardless of what specific numbers we use, provided that **a** represents a positive number and **x** and **m** represent any real numbers. Write out the whole argument on your paper, filling in the blanks as you go to complete it..

Given	$\log (a)^m = x$
Change to exponent form	_____
Raise both sides to the $^1/_m$ power	_____
Change back to log form	_____
_____	$m \log a = x$
Substituting for x	$\log (a)^m = m \log a$

Be sure to add a statement describing this property of logarithms to your Tool Kit along with at least three numerical examples.

CC-110.   How can we use this idea to solve your old friend $2 = 1.04^x$? If two expressions are equal would their logarithms, base 10 be equal? Solve for x, accurate to eight decimal places.

CC-111.   Use the ideas from CC-108-110 to solve this equation. Be accurate to three decimal places.

$$5 = 1.04^x$$

CC-112.   Solve each of the following (nearest 0.001):

a) $25^x = 145$

b) $(1.28)^x = 4.552$

c) $240(0.95)^x = 100$

CC-113. Use the idea from CC-109 to write three different looking equivalent expressions for each of the following. For example: $\log 7^{3/2}$ can be written as $\frac{3}{2} \log 7$, $\frac{1}{2} \log 7^3$, $3\log \sqrt{7}$, etc.

a) $\log 8^{2/3}$ b) $-2 \log 5$ c) $\log (na)^{ba}$

CC-114. Margee thinks she can use logs to solve $56 = x^8$ since logs seem to make exponents disappear. Unfortunately, Margee is wrong. Explain the difference between equations like $2 = 1.04^x$, where you can use logs, and $56 = x^8$ where you don't need logs.

CC-115. Investigate:

a)   What can we multiply 8 by to get 1?

b)   What can we multiply **a** by to get 1?

c)   By using the rules of exponents, find a way to solve $m^8 = 40$ without having to use logarithms. (Obtain the answer as a decimal approximation using your calculator. Check your result by raising it to the 8th power.)

d)   Now solve $n^6 = 300$.

e)   Find a rule for solving $x^a = b$ for x with a calculator.

CC-116. What is the equation of the line of symmetry of the graph of $y = (x - 17)^2$? Justify your answer.

CC-117. Solve each system below:

a) $-4x = z - 2y + 12$
      $y + z = 12 - x$
      $8x - 3y + 4z = 1$

b) $3x + y - 2z = 6$
      $x + 2y + z = 7$
      $6x + 2y - 4z = 12$

c)   What does the solution in part (b) tell you about the graphs?

CC-118. Write as powers of x:

a) $\sqrt[5]{x}$ b) $\dfrac{1}{x^3}$

c) $\sqrt[3]{x^2}$ d) $\dfrac{1}{\sqrt{x}}$

CC-119   Below is a graph of $y = \log_b x$. Find a reasonable value for b.

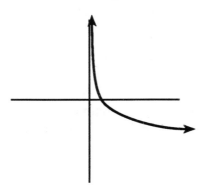

CC-120   Describe the transformation and sketch the graph of $y = \log_3(x + 4)$.

WRITING EQUATIONS FOR EXPONENTIAL FUNCTIONS

CC-121.   As you learned in Chapter 3, an exponential function has the general form:

$$y = am^x.$$

a)   Does this type of equation have an asymptote? If so, what?

b)   A particular exponential function goes through the points (2, 36) and (1, 12), and has the x axis as a horizontal asymptote. Substitute these values into $y = am^x$, for x and y to create two equations with two unknowns, the unknowns being **a** and **m**.

c)   Solve one of your equations for **a** or **m**. Substitute that result into the remaining equation. You should now be able to figure out one of the variables. Once you know one of the variables, you should be able to find the other.

d)   What is the exponential function that passes through (2, 36) and (1, 12)?

CC-122.   I'd like to have $40,000 in 8 years (hey, I'm greedy), and I only have $1000 now.

a)   What interest rate do I need to find which, when compounded yearly, will fill my needs?

To help me solve this, set up an exponential equation: let y = amount of money, and let x = number of years, and let m be the multiplier. Since interest rate problems are best modeled by exponential functions, the equation is to be in the form $y = km^x$.

b)   Try another one: I start with $7800 and I want to have $18,400 twenty years from now. What interest rate do I need (compounded yearly)?

c)   Which of these two scenarios do you think is more likely to happen? Justify your response.

CC-123.   An exponential function contains these two points: (3, 12.5) and (4, 11.25).

    a)   Is the exponential function increasing or decreasing? Justify your answer.

    b)   This exponential function does not have the x-axis as a horizontal asymptote. The horizontal asymptote for this function is the line y = 10. Make a sketch of this graph showing the horizontal asymptote.

    c)   If this function has the equation $y = km^x + b$, what would be the value of b? Explain.

    d)   Substitute the known points into the equation $y = km^x + 10$ and solve the system for k and m.

    e)   What is the equation of the function?

CC-124.   Solve each of the following (to the nearest 0.001).

    a) $(5.825)^{x-3} = 120$ b) $18(1.2)^{2x-1} = 900$

CC-125.   **HOLLYWOOD GLITTER**

The economy has worsened to the point that the merchants in downtown Hollywood cannot afford to replace the light bulbs when they burn out. On average about thirteen percent of the light bulbs burn out every month. Assuming there are now about one million outside store lights in Hollywood, how long will it take until there are only 100,000 bulbs lit? Until there is only one bulb lit?

CC- 126.   Using your calculator, compare each expression in Column A with its paired expression in Column B.

$$\begin{matrix} > \\ < \\ = \\ ? \end{matrix}$$

Column A	?	Column B
log 30		log 5 + log 6
log 27		log 9 + log 3
log 24		log 2 + log 12
log 132		log 12 + log 11

    a)   Write a conjecture (a statement you think is true) based on the pattern you noticed.

    b)   What is another way of expressing log 65?

    c)   Make up three more examples of your own, and check them.

CC-127.   Using your calculator, compare the expressions in these two columns.

$$\underline{\text{Column C}} \qquad\qquad \underline{\text{Column D}}$$

Column C	Column D
log 30	log 300 - log 10
log 27	log 81 - log 3
log 8	log 24 - log 3
log 12	log 60 - log 5

a)   Write a conjecture based on the pattern you noticed.

b)   What is another way of expressing log 13?

c)   Make up three more examples of your own, and check them.

CC-128.   (Last time!) Using your calculator, compare the expressions in these two columns.

$$\underline{\text{Column E}} \qquad\qquad \underline{\text{Column F}}$$

Column E	Column F
$\log(3 \cdot 4)$	$\log 3 + \log 4$
$\log\left(\frac{72}{8}\right)$	$\log 72 - \log 8$

a)   What is another way of expressing $\log (a \cdot b)$?

b)   What is another way of expressing $\log \left(\frac{a}{b}\right)$?

CC-129.   Solve each system of equations:

a) $2x + y = 1$
     $-3y = 10x$

b) $\dfrac{x + y}{5} = 1$

     $\dfrac{2x + 3y}{3} = 1$

CC-130.   Kirsta was working with her function machine, but when she turned her back her brother Caleb dropped a number in. She didn't see what was dropped in, but she did see what fell out: 9.

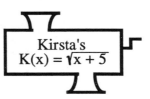

Kirsta's
$K(x) = \sqrt{x + 5}$

a)   What operations must she perform on 9 to undo what her machine did? Use this to find out what Caleb dropped in.

b)   Write a rule for a machine that will undo Kirsta's machine. Call it c(x).

CC-131.   Is the graph at right a function? Explain

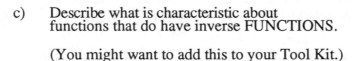

   a)   Make a sketch of the inverse of this graph. Is
         the inverse a function? Justify your answers.

   b)   Must the inverse of a function be a function?
         Explain.

   c)   Describe what is characteristic about
         functions that do have inverse FUNCTIONS.

         (You might want to add this to your Tool Kit.)

   d)   Could the inverse of a non-function be a function? Explain or give an example.

CC-132.   The half-life of uranium is 1000 years. If 50 grams of uranium is sealed in a box,

   a)   how much is left after 10,000 years?

   b)   how long will it take to reduce to 1% of the original amount?

   c)   how long will it take until all of the original mass of uranium is gone. Support
         your answer.

CC-133.   **THE CASE OF THE COOLING CORPSE**

The coroner's office is kept at a cool 17°C. Agent 008 kept pacing back and forth
trying to keep warm as he waited for any new information. For over three hours now,
Dr. Dedman had been performing an autopsy on the Sideroad Slasher's latest victim,
and Agent 008 could see that the temperature of the room and the deafening silence
were beginning to irritate even Dr. Dedman. The slasher had been creating more work
than Dr. Dedman cared to investigate.

"Dr. Dedman, don't you need to take a break? You've been examining this dead body
for hours! Even if there were any clues, you probably wouldn't see them at this point."
Agent 008 queried.

"I don't know." Dr. Dedman replied. "I just have this feeling something is not quite
right. Somehow the slasher slipped up with this one and left a clue. We just have to
find it."

"Well, I have to check in with HQ," 008 stated. "Do you mind if I step out for a
couple of hours?"

"No, that's fine." Dr. Dedman responded. "Maybe I'll have something by the time
you return."

Yeah, right, 008 thought to himself. These small town people always want to be the
hero and solve everything. They just don't realize how big this case really is. The
Slasher has left a trail of dead bodies through five states! 008 left, closing the door
quietly. As he walked down the hall, he could hear the doctor's voice fade away, as he
described the victim's gruesome appearance into the tape recorder.

>>>**PROBLEM CONTINUES ON NEXT PAGE**>>>

The hallway from the coroner's office to the elevator was long and dark. This was the only way to Dr. Dedman's office. Didn't this frighten most people? Well, it didn't seem to bother old Ajax Boraxo who was busy mopping the floor, thought 008 . . . most of the others wouldn't notice, he reminded himself.

He stopped briefly to use the restroom and bumped into one of the deputy coroners. "Dedman still at it?" "Sure is, Dr. Quincy. He's totally obsessed. He's dead sure there is a clue." As usual, when leaving the court house, 008 had to sign out. Why couldn't these small towns move into the 20th century? **Everyone** used surveillance cameras these days.

"How's it going down there Agent?" Sergeant Foust asked. Foust spent most of his shifts monitoring the front door, forcing all visitors to sign-in, while he recorded the time next to the signature. Agent 008 wondered if Foust longed for a more exciting aspect of law enforcement. He thought if he were doing Foust's job, he would get a little stir-crazy sitting behind a desk most of the day. Why would someone become a cop to do this?

"Dr. Dedman is convinced he will find something soon. We'll see!" Agent 008 responded. He noticed the time: Ten minutes to 2:00. Would he make it to HQ before the chief left?

"Well, good luck!" Foust shouted as 008 headed out the door.

008 sighed deeply when he returned to the court house. Foust gave his usual greeting: "Would the secret guest please sign in!" he would say, handing a pen to 008 as he walked through the door. Sign in again, he thought to himself. This takes so long. 5:05 p.m. Agent 008 had not planned to be gone so long, but he had been caught up in what the staff at HQ had discovered about that calculator he had found. For a moment he saw a positive point for having anyone who comes in or out of the court house sign in: he knew by quickly scanning the list that Dr. Dedman had not left. In fact, the old guy must still be working on the case.

As he approached the coroner's office, he had a strange feeling that something was wrong. He couldn't hear or see Dr. Dedman. When he slowly opened the door, the sight he saw inside stopped him in his tracks. Evidently, Dr. Dedman had become the newest victim of the slasher. But wait? The other body, the one the doctor had been working on, was gone! Immediately, the security desk, with its annoying sign-in sheet came to mind. Yes there were lots of names on that list, but if he could determine the time of Dr. Dedman's death, he might be able to just scan the roster to find the murderer! Quickly, he grabbed the thermometer to measure the Doctor's body temperature. He turned around and hit the security buzzer. The bells were deafening. He knew the building would be sealed off instantly and security would be there within seconds.

"My God!" Foust cried as he rushed in. "How did this happen? Who could have done this? I spoke to the Doctor less than an hour ago. What a travesty!"

As the security officers crowded into the room, Agent 008 explained what he knew, which was almost nothing. He stopped long enough to check the doctor's body temperature: 27°C - that's 10°C below normal. He figured that the doctor had been dead at least an hour. Then he remembered: the tape recorder! Dr. Dedman had been taping his observations - that was standard procedure. They began looking everywhere

>>>PROBLEM CONTINUES ON NEXT PAGE>>>

for that blasted thing. Yes, the slasher must have realized that the doctor had been taping and taken that as well. Exactly an hour had passed during the search and Agent 008 noticed that the thermometer still remained in Dr. Dedman's side. The thermometer clearly read 24°C. Agent 008 knew he could now determine the time of death precisely.

Coroner's Office - Please Sign In		
Name	Time In	Time Out
Rufin Vonboskin	12:08	2:47
Lizzy Borden	12:22	1:38
Chuck Manson	12:30	2:45
Hanibal Lechter	12:51	1:25
Ajax Boraxo	1:00	2:30
D. C. Quincy	1:10	2:45
Agent 008	1:30	1:50
Ronda Ripley	1:43	2:10
Jeff Domer	2:08	2:48
Stacy Stiletto	2:14	2:51
Blade Butcher	2:20	2:43
Pierce Slaughter	3:48	4:18
Gashes Wound	3:52	5:00
Slippery Eel	3:57	4:45
Candy Carcass	4:08	4:23
Milly Maniacal	4:17	4:39
D.C. Quincy	4:26	4:50
Fred Cruger	4:35	
Danny Demented	4:48	4:57
Larry Laceration	5:04	
Agent 008	5:05	
Security	5:12	

a) Make a sketch showing the relationship between body temperature and time. What type of function is it? Justify your answer.

b) What is the asymptote for this relationship? Explain.

c) Use your data and the equation $y = km^x + b$ to find the equation that represents the temperature of the body at a certain time.

d) When did Dr. Dedman die?

e) Who is the murderer?

CC-134.   **CHAPTER 6 SUMMARY ASSIGNMENT**

a)   Throughout the year so far, you have been working with different groups. What are the types of contributions you have made? What role have you taken in your group? Would you like these to change or stay the same? Explain.

b)   What are the important topics of this chapter?

c)   For each of the topics you chose in part (b), find a problem that represents that topic and that you enjoyed doing. Write out the problem (or outline the key pieces of information of the problem) along with the complete solution.

d)   Find a problem that you still cannot solve. Write out the problem and as much of the solution as you can. What is it you need to know to finish the problem?

e)   How does this chapter seem to connect to any other topics in mathematics that you have studied? Explain.

CC-135.   A rule-of-thumb used by car dealers is that the trade-in value of a car decreases by 30% each year.

a)   Explain how the phrase "decreases by 30% each year" tells you that the trade-in value varies **exponentially** with time (i.e. can be represented by an exponential function).

b)   Suppose the initial value of the car is $2350. Write an equation expressing the trade-in value of your car as a function of the number of years from the present.

c)   How much is the car worth in three years?

d)   In how many years will the trade-in value be $600?

e)   If the car is really 2.7 years old now, what was its trade-in value when it was new?

CC-136.   **PORTFOLIO: GROWTH OVER TIME PROBLEM**

On a separate sheet of paper, explain **everything** that you now know about:

$$f(x) = x^2 - 4 \quad \text{and} \quad g(x) = \sqrt{x + 4}$$

CC-137.   **GROWTH OVER TIME REFLECTION**

Look at your three responses to the growth-over-time problem. Write a self evaluation of your growth based those responses. Consider the following while writing your answer:

a)   What new concepts did you include the second time you did the problem? In what ways was your response better than your first attempt?

b)   How was your final version different from the first two? What new concepts did you include?

c)   Did you omit anything in the final version that you used in one of the earlier problems? Why did you omit that item?

>>>PROBLEM CONTINUES ON NEXT PAGE>>>

d)   Draw some bars like the ones below and shade the amount that you knew when you did the problem at each stage. (blank means you knew nothing about the problem up to completely shaded meaning you knew and answered everything that could be answered.)

First Attempt: [   ]

Second Attempt: [   ]

Third Attempt: [   ]

e)   Is there anything you would add to your most recent version? What?

CC-138. a)   Explain completely how to get a good sketch of the graph of $y = (x + 6)^2 - 7$.

b)   Explain how to change the graph to now be the graph of $y < (x + 6)^2 - 7$.

c)   Given the original graph, how can you get the graph of $y = |(x + 6)^2 - 7|$.

d)   Restrict the domain of the original parabola to $x \geq -6$, and graph the inverse function.

And for those who need a challenge, ...

e)   What would be the inverse function if we had restricted the domain to $x \leq -6$?

CC-139. Solve for x. (Do it without a calculator.)

a) $x = \log_{25}(5)$ b) $\log_x(1) = 0$ c) $23 = \log_{10}(x)$

CC-140. Solve for a positive number to the nearest 0.001, on your calculator:

a) $x^6 = 125$ b) $x^{3.8} = 240$ c) $x^{-4} = 100$

d) $(x + 2)^3 = 65$ e) $4(x - 2)^{12.5} = 2486$ f) $n^3 = 49$.

CC-141. Find the inverse of each of the functions below. Write your answers in function notation.

a) $p(x) = 3(x^3 + 6)$ b) $k(x) = 3x^3 + 6$

c) $h(x) = \dfrac{x + 1}{x - 1}$ d) $y = \dfrac{2}{3 - x}$

CC-142. A two bedroom house in Vacaville is worth $110,000. If it appreciates at a rate of 2.5% each year,

a)   what will it be worth in 10 years?

b)   when will it be worth $200,000?

c)   In Davis, houses are depreciating at a rate of 5% each year. If a house is worth $182,500 now, how much would it be worth two years from now?

MATHEMATICS
REPORT ON GROWTH

We have been learning about several different families of functions such as lines, absolute value, parabolas, exponential, and others. All of these can be used to describe something in the real world. In this project, you will learn more about how mathematical ideas are used to model reality. This project has four parts.

1.   **Project Proposal:** The first step is to choose something that has been growing or shrinking steadily over time. Be creative! Choose something you are really interested in studying: cancer rates, world record times for some sport, crime rates, AIDS, teenage pregnancy rates, the number of lawyers in the US, or the population of your family's country of origin are some possibilities. You are free to choose one of these or a topic of your own.
     Once you have chosen your topic you must turn in a proposal. In your proposal, you must state your topic, explain why you chose it, and describe exactly where you plan to collect your data. **Don't** say "at the library!" Describe in detail exactly where: who are you going to call or ask? In what book will you look? These must be specific as this is your bibliography for your project. Note that your data must show an increase or decrease over the last **three** time periods (days, hours, months, years - whatever is relevant for the topic).

2.   **Mathematical Modelling:** Your job in this part is to construct different mathematical models for your data, and use the models to make predictions about the future. Using your three or more data points, find the equation of a line, a parabola, and an exponential function that best fit the data. For extra credit, you may use other families of functions to fit your data. Make a large graph (full sheet of paper), clearly labelled, for each. Show **completely** how you derived your equations (this is one of the most important parts).

3.   **Analysis:** Finally, tell what your equations predict for the future and explain what the consequences of your predictions are. Which model do you think is best, and why? What factors would make your model less reliable? Is there some point at which your model is obviously wrong? You must be complete.

4.   **Report Presentation:** The report should be typed, neatly presented, and complete.

The time line and point values are as follows:

TOPIC	DUE DATE	% VALUE
Proposal		
Modelling		
Analysis		
Presentation		

This project is to be done **individually**. All written material must be typed. Please note all due dates. Be sure to include **at least** everything mentioned above.

LOGARITHMS TOOL KIT

Logs and Exponents:
    (conversions)

Multiplication/Addition

Division/Subtraction

Powers/Coefficients

Change of Base

Calculator Buttons

Natural Log (ln and "e")

Chapter 7
Polynomials and General Systems

(At the County Fair)

PROBLEM SOLVING

REPRESENTATION/MODELING

FUNCTIONS/GRAPHING

INTERSECTIONS/SYSTEMS

ALGORITHMS

REASONING/COMMUNICATION

CHAPTER 7
AT THE COUNTY FAIR:
POLYNOMIALS AND GENERAL SYSTEMS

In this chapter you will have the opportunity to:

- Explore a new category of equations and their graphs: polynomials.

- Find intersections of non-linear functions and other equations both graphically and algebraically.

- Discover that some systems of equations cannot be solved algebraically. Their solutions can only be approximated by graphing and estimating.

- Learn that other equations which, at first glance, could not be solved can be solved when we create a whole new set of numbers.

- Expand your thinking in order to work with numbers composed of real and imaginary parts.

- Summarize the important ideas from the course.

Keep a graphing calculator handy - you will be surprised at how useful, even necessary, it will be in solving many of these problems!

CF-0.   **THE GAME TANK**

The Mathamericaland Carnival Company wants to make a special game to premiere at this year's county fair. The game will consist of a tank filled with ping pong balls. Most are ordinary white ones, but there are a limited number of orange, red, blue, and green prize ping pong balls. The blue ones have a prize value of $2.00; the reds, $5.00; the orange ones, $20.00; and the green ping pong ball (there's only one!) is worth $1000.00. There are 1000 blue ping pong balls, 500 reds, and 100 orange ones. Fairgoers will pay $1.00 for the opportunity to crawl around in the tank for a set amount of time, blindfolded, trying to find one of the colored ping pong balls.

The owner of the company thinks that she will make the most profit if the tank has maximum volume. In order to cut down on the labor costs of building this new game, she hires you to find out what shape to make the tank.

POLYNOMIAL FUNCTIONS LAB
(CF-1 through CF-3)

CF-1.   We are going to be doing an investigation of polynomial functions, but first we will do one example as a class. There is a resource page for these problems at the end of the chapter. The example function we will work with is $P_1(x) = (x - 2)(x+ 5)^2$.

   a)   Graph the function on the graphing calculator using the standard viewing window.

   b)   As you can see we do not see the best view of the function, we want to use the zoom features of the graphing calculator to obtain a better view of the graph. What should we be looking for when we zoom? Start by zooming out. You may have to do this more than once to get the complete graph. Now use the box feature to select the portion of the graph which you would like to view. Sketch this graph on the second set of axes. It helps to label the important points of the graph such as the intercepts.

   c)   Find the roots. The roots are the solutions of the equation $0 = (x - 2)(x+ 5)^2$

   d)   On the number line mark the roots with open circles and then shade the regions where the function outputs are positive (the graph is above the x-axis).

   e)   Describe the graph and its relation to the equation. Make sure that you include the degree and the intercepts as part of your description. Pay close attention to the way the function curves.

⚐CF-2.   In the Parabola Lab in chapter 4, you discovered how to make a parabola "sit" on the x-axis, and you also looked at ways of making parabolas intersect the x-axis in two specific places. These x-intercepts for the graph of the function are often called the **roots of the function** and sometimes can be found by factoring. Add **roots** to your Tool Kit.

CF-3.   Use the approach you used for $P_1(x)$ to investigate the following five polynomials.

   a) $P_2(x) = (x - 1)(x + 1)(x - 3)$ b) $P_3(x) = 0.2x(x + 1)(x - 3)(x + 4)$

   c) $P_4(x) = (x + 3)^2(x + 1)(x - 1)$ d) $P_5(x) = -0.1x(x + 4)^3$

   e) $P_6(x) = x^4 - x^2$

CF-4.   Find the **roots** of each of the following functions:

   a) $y = x^2 - 6x + 8$ b) $f(x) = x^2 - 6x + 9$ c) $y = x^3 - 4x$

CF-5.   Graph: $y = (x - 1)^2(x + 1)$. Make a table from -2 to 2.

   a)   What do you think is the parent for this equation?

   Now graph: $y = (x - 1)^2(x + 1)^2$. Again, a table from -2 to 2 would be appropriate.

   b)   What do you think is the parent for this equation?

   Finally, sketch a graph of $y = x^3 - 4x$.

CF-6.   Algebraic expressions, like those in the previous problem, that include at most addition, subtraction and multiplication are called **polynomials**.

These are polynomial functions	These are not polynomial functions
a) $f(x) = 8x^5 + x^2 + 6.5x^4 + 6$	d) $y = 2^x + 8$
b) $y = \frac{3}{5}x^6 + 19x^2$	e) $f(x) = 9 + \sqrt{x} - 3$
c) $P(x) = 7(x - 3)(x + 5)^2$	f) $y = x^2 + \dfrac{1}{x^2 + 5}$

The general expression for a polynomial with one variable is often written in textbooks as:
$$(a_n)x^n + (a_{n-1})x^{n-1} + \ldots + (a_1)x^1 + (a_0)$$

Where n is a positive integer and $a_n, a_{n-1}, \ldots, a_1, a_0$ usually represent specific numbers. For example:

in $7x^4 - 5x^3 + 3x^2 + 7x + 8$,   n = 4, therefore . . .

$a_n = a_4 = 7$, $a_{n-1} = a_3 = -5$, $a_{n-2} = a_2 = 3$, $a_1 = 7$ & $a_0 = 8$

QUESTION: Why are (d), (e) and (f) <u>not</u> polynomial functions?

CF-7.   In the previous problem, polynomial function (a) has degree 5, polynomial function (b) has degree 6 and, polynomial function (c) has degree 3. Figure out the meaning of the term **degree** of a polynomial function and explain how you can tell (c) has degree 3. Discuss this with your group and include this definition in your Tool Kit along with your own definition of a polynomial function.

CF-8.   Describe the possible numbers of intersections for each of the following pairs of graphs. Sketch a graph for each possibility. For example, in part (b), a parabola could intersect a line twice, once, or not at all. Your solution to each part should include all of the possibilities and a sketched example of each one.

a)   Two different lines.

b)   A line and a parabola.

c)   Two different parabolas.

d)   A parabola and a circle.

CF-9.   Solve the following system: $y = x^2 - 5$
                                                           $y = x + 1$

CF-10.   In the previous problem:

a)   what method did you use?

b)   can this problem be solved algebraically? Explain.

CF-11. Which of the following are polynomial functions? For each one that is not, give a brief justification as to why it is not.

a) $y = 3x^3 + 2x^2 + x$

b) $y = (x - 1)^2(x - 2)^2$

c) $y = x^2 + 2^x$

d) $y = 3x - 1$

e) $y = (x - 2)^2 - 1$

f) $y^2 = (x - 2)^2 - 1$

g) $y = \dfrac{1}{x^2} + \dfrac{1}{x} + \dfrac{1}{2}$

h) $y = \dfrac{1}{2}x + \dfrac{1}{3}$

CF-12. Describe the difference between the graphs of $y = x^3 - x$ and $y = x^3 - x + 5$.

CF-13. Tamar thinks that the equation $(x - 4)^2 + (y - 3)^2 = 25$ is equivalent to the equation $(x - 4) + (y - 3) = 5$, because you can just take the square root of both sides of the first equation. Are the two equations equivalent? Explain why or why not.

MORE ANALYSIS OF POLYNOMIAL FUNCTIONS

CF-14. Look back at the work you did on the **Polynomial Functions Lab**. Then answer the following:

a) What is the maximum number of roots a polynomial of degree 3 can have? Sketch an example.

b) What is the maximum number of roots a polynomial of degree **n** can have?

c) Can a polynomial of degree **n** have fewer than **n** roots? Under what conditions?

d) Match each function with the minimum degree its polynomial equation could have:

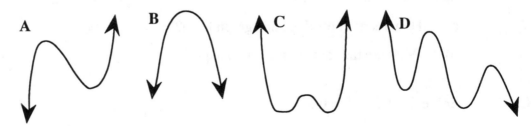

e) Which of the graphs above have a negative orientation? (They are opposite of their parent graphs.) Explain how you determine the orientation of a graph.

CF-15. In the previous problem, we specified how to find the minimum degree. It may be possible that these graphs have a higher degree.

Look at the graphs of $y = x^2$ and $y = x^4$.

a) How are these graphs similar? How are they different?

b) Could **B** (previous problem) have degree 4?

c) Could it have degree 5? Explain.

CF-16. In the first example from the **Polynomial Functions Lab**, $(x + 5)^2$ is a factor. This produces what is called a **double root**.

a)   What affect does this have on the graph?

b)   In $P_5(x)$, there is a **triple root**. What does the equation have that lets us know it has a "triple root?"

c)   What does this do to the graph?

CF-17. The following number lines show where the output values of each polynomial function (a), (b), (c), & (d) are positive, in other words, where the graph is above the x-axis. Sketch a possible graph to fit each description.

a)

b)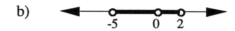

c)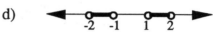

d)

(Note: Each number line represents a different polynomial.)

CF-18. Sketch rough graphs of the following functions:

a) $P(x) = -x(x + 1)(x - 3)$

b) $P(x) = (x - 1)^2(x + 2)(x - 4)$

c) $P(x) = (x + 2)^3(x - 4)$

CF-19. Where does the graph $y = (x + 3)^2 - 5$ cross the x-axis?

CF-20. Sketch a graph of $y = x^2 - 7$

a)   How many roots does this graph have?

b)   What are the roots of the function?

CF-21. Solve: $x^2 + 2x - 5 = 0$.

a)   How many roots does $y = x^2 + 2x - 5$ have?

b)   What are they?

c)   Approximately where does the graph of $y = x^2 + 2x - 5$ cross the x-axis?

CF-22. Are parabolas polynomial functions? Explain.

Are lines polynomial functions? Explain.

How about cubics? Exponentials? Circles? Explain, explain, explain . . .

CF-23.   If you were to graph the function $f(x) = (x - 74)^2(x + 29)$, where would the graph of $f(x)$ intersect the x-axis?

CF-24.   What is the degree of each of these polynomial functions.

   a) $P(x) = 0.08x^2 + 28x$

   b) $y = 8x^2 - \frac{1}{7}x^5 + 9$

   c) $f(x) = 5(x + 3)(x - 2)(x + 7)$

   d) $y = (x - 3)^2(x + 1)(x^3 + 1)$

CF-25.   Graph the inequality $x^2 + y^2 \leq 25$, then describe its graph in words.

CF-26.   Find x if $2^{p(x)} = 4$ where $p(x) = x^2 - 4x - 3$.

CF-27.   Start with the graph of $y = 3^x$, then write new equations that will shift the graph . . .

   a)   down 4 units.

   b)   to the right 7 units.

CF-28.   Explain what the relationship is between the solutions of $3x = x^2 + 5x$ and the x values where the graphs of $y = 3x$ and $y = x^2 + 5x$ intersect.

CF-29.   Judy claims that since $(xy)^4$ is x^4y^4, it must be true that $(x + y)^4$ is $x^4 + y^4$ What do you think? Explain your reasoning

NON-LINEAR SYSTEMS OF EQUATIONS

CF-30.   Consider the equation $2^x = x + 3$.

   a)   Discuss in your group different methods you could use to solve this equation.

   b)   Solve the equation.

   c)   Be sure you found two solutions. How can you be certain there are no more than two?

CF-31.   When solving systems of equations, it is important to recall the significance of the solution. What does the solution represent in terms of . . .

   a)   the equations?

   b)   the graphs of the equations?

CF-32.   Consider the following system: $y = 2^x + 1$
                                                 $y = x + 6$

   a)   What kind of equation is $y = 2^x + 1$?

   b)   What kind of equation is $y = x + 6$?

   c)   In how many points could these graphs intersect? Must they intersect? (This will
         tell us how many solutions we can expect)

   d)   Solve the system by any method that your group discussed for the previous problem.

CF-33.   Kenya needs to graph $y = \log_2 x$ and wants to use her graphing calculator. She figures
          that she can change the equation into exponential form and then re-solve for y. Here is
          what she did:

$$y = \log_2 x$$

$$2^y = x$$

$$\log(2^y) = \log(x) \qquad\qquad \text{(remember, this is}$$
$$\text{now log base 10.)}$$

$$y \cdot \log(2) = \log(x)$$

$$y = \frac{\log(x)}{\log(2)}$$

   a)   Will this work for any base? Show the method for $y = \log_7 x$.

   b)   How would you change $y = \log_3(x + 5)$ to be able to graph it on a graphing
         calculator?

   c)   How would you change $y = \log_2 x - 3$ to be able to graph it on a graphing calculator?

CF-34.   Find where the graphs of $y = \log_2(x - 1)$ and $y = x^3 - 4x$ intersect.

CF-35.   Solve $\log_2(x - 1) = x^3 - 4x$.

CF-36.   The systems that you have been solving so far today are non-linear systems.

   a)   What does "non-linear" mean?

   Some non-linear systems may be **impossible** to solve algebraically; others may just be
   extremely difficult.

   b)   What other methods do we have to solve these systems?

CF-37.   Sketch a graph of the equation $y = x^2 + 4$:

a)   Where does the graph cross the x-axis?

b)   Solve the equation $x^2 + 4 = 0$ and explain how this relates to the answer you found in part b).

c)   What occurred when you tried to solve the equation in part (b) that let you know that it has no real solution?

CF-38.   Solve the following equation for **x**:

a) $\dfrac{x + 3}{x - 1} - \dfrac{x}{x + 1} = \dfrac{8}{x^2 - 1}$

b)   In part (a) the result of solving the equation is x = 1, but what happens when you substitute 1 for x? What does this mean in relation to solutions for this equation?

CF-39.   A sequence of pentagonal numbers is started to the right.

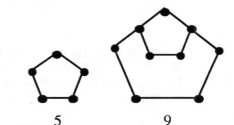

a)   Find the next 3 pentagonal numbers.

b)   What kind of a sequence is formed by the pentagonal numbers?

c)   What is the equation for the **n**th pentagonal number?

1   5   9

CF-40.   A circle with its center on the line y = 3x (1st quadrant) and is tangent to the y-axis.

a)   If the radius is 2, what is the equation of the circle?

b)   If the radius is 3, what is the equation of the circle?

CF-41.   Sketch on the same set of axis:

a) $y = 2^x$ b) $y = 2^x + 5$ c) $y = 2^x - 5$

CF-42.   Sketch the graphs and find the area of the intersection of:

$$y > |x + 3|$$
$$y \le 5$$

CF-43.   Verify that x = -2 is a solution to the equation, $x^4 - 4x = 8x^2 - 40$

CF-44.   Solve: (remember "fraction busters?")

a) $\dfrac{3x}{x + 2} + \dfrac{7}{x - 2} = 3$

b) $\dfrac{x - 7}{x - 5} - \dfrac{6}{x} = 1$

CF-45.   An arithmetic sequence starts out -23, -19, -15, . . .

a)   What is the rule?

b)   How many times must the initial value be put through the generator so that the result is greater than 10,000?

CF-46.   A contractor, who was working for the county, failed to complete the new County Fair Pavilion within a specified time. According to his contract, he is compelled to forfeit $10,000 a day for the first ten days of extra time required, and for each additional day, beginning with the eleventh, the forfeit is increased by $1000 a day. If he lost a total of $255,000, how many days did he overrun the stipulated completion date? What kind of sequence do his fines form starting after the tenth day?

CF-47.   Artemus was putting up the sign at the County Fair Theater for the movie "ELVIS RETURNS FROM MARS," (a sure draw for this year's fair!). He got all of the letters he would need and put them in a box. He reached into the box and pulled out a letter at random.

a)   What is the probability that he got the first letter he needed when he reached into the box?

b)   Once he put the first letter up, what is the probability that he got the second letter he needed when he reached into the box?

MORE NON-LINEAR SYSTEMS

CF-48.   Sketch the circle $(x + 2)^2 + (y + 1)^2 = 9$ and the line $y = x + 4$ on the same set of axes.

Find and clearly label the point(s) of intersection of the line and the circle.

CF-49.   Sketch the circle $x^2 + y^2 = 25$, and the parabola $y = x^2 - 13$.

a)   How many points of intersection are there for this parabola and the circle?

b)   Find the coordinates of these points algebraically. Remember the key to solving systems is to eliminate a variable.

c)   Explain how the graph helps you with the algebraic solution.

CF-50.   Solve this system of equations, and explain your method:

$$(x - 1)^2 + y^2 = 16$$
$$y = 2^x$$

CF-51.   Solve this system of equations. Be sure to sketch the graphs on your paper.

$$(x - 2)^2 + y^2 = 25$$
$$y = x^3 - 9x$$

a)   How many solutions does this system have? Are you sure?

b)   Zoom in or set up a new range with x-values close to -3, and y-values a little greater than zero. Now what do you think?

CF-52.   Verify that the graphs of the equations $x^2 + y^3 = 17$ and $x^4 - 4y^2 - 8xy = 17$ intersect at (3, 2). Then:

a)   give an equation of the vertical line through this point.

b)   give an equation of the horizontal line through this point.

CF-53.   Each year the Strongberg Construction company builds twenty more houses than in the previous year. Last year they built 180 homes. Acme Homes business is increasing by 15% each year and they built 80 homes last year.

a)   Assuming that you don't have a graphing calculator with you now, what can you do to solve this problem with just a scientific calculator?

b)   Write equations for each construction company and find the year in which both construction companies will build the same number of homes.

CF-54.   Each of the dart boards below is a target at the County Fair dart throwing game. What is the probability of hitting the **shaded** region of each target (assuming you always hit the board but where on the board is random)?

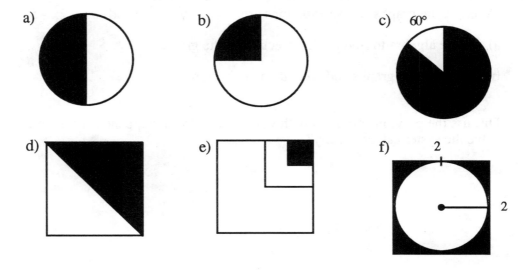

CF-55.   Julianna says that $\dfrac{a}{x+b} = \dfrac{a}{x} + \dfrac{a}{b}$. Fred is not sure; he thinks that the two expressions are not equivalent. Who is correct? Justify your answer.

CF-56.   The area of $\triangle ABC$ is 24 square inches. If $\overline{AB} \perp \overline{BC}$ and $BC = 8$ in., find AC.

CF-57.   Solve: $y^2 + 2y = 1$.

CF-58.   A general equation is an equation that can be used to describe a group of functions with similar characteristics (sometimes called a family of functions.). For example, circles with centers at (-3, 5) can be written as $(x + 3)^2 + (y - 5)^2 = k^2$. Here k can be any positive number. Find the general equation of each of the following and sketch a couple of examples of each type:

a)   Lines parallel to $y = 6x - 7$

b)   All parabolas with a vertex at (4, 3).

c)   All parabolas with roots of 2 and -5.

CF-59.   Choosing your own values for k, sketch on the same set of axes five different examples of graphs which fit the equation: $(x - 3)^2 + (y - 5)^2 = k$. Then describe the graphs in words.

CF-60.   Choosing different values for k and j, sketch three different examples of graphs which fit the equation: $(x - k)^2 + (y - j)^2 = 9$. Then describe the graphs in words.

CF-61.   Sketch three different graphs which fit the equation: $(x - d)^2 + (y - 5)^2 = 16$ by choosing different values of d. Describe the graphs in words.

A NEW KIND OF NUMBER

CF-62.   Consider the graphs whose equations are: $y = x^2$ and $y = 2x - 5$

a)   Use algebra to find the intersection of the graphs.

b)   Sketch the graphs and label the intersection.

CF-63.   Discuss the previous problem with your group. What happened when you tried to solve the system algebraically?

CF-64.   Solve: $x^2 = 2$.

COUNTY FAIR INFORMATION BOOTH

In Ancient Greece, people believed that all numbers could be written as fractions (one integer over another). Many individuals realized later that some numbers could not be written as fractions (such as the result you just calculated) and challenged the accepted beliefs. Some of these people were exiled or outright killed over these challenges. As you can see, the Greeks took their mathematics very seriously. Eventually, they could not avoid the situation and invented a new way of thinking. They even invented new numbers, square roots. This problem has no rational solutions. Fractions won't work, but square roots will. There are decimal approximations that can be found on your calculator, but the exact answer can only be represented in radical form.

---

a)   How do you "undo" square?

b)   When you solved $x^2 = 2$, how many solutions did you get?

c)   How many x-intercepts will $y = x^2 - 2$ have?

d)   Write your solutions both as radicals and as decimal approximations.

CF-65.   Mathematicians throughout history have challenged the notion that some equations were not solvable and have invented new ways of dealing with these situations. Recently you have come across a similar situation where a solution to a problem could not be found because of an existing mathematical belief.

a)   What happens if we try to solve $x^2 = -1$?

COUNTY FAIR INFORMATION BOOTH (part ii)

Instead of simply settling on no solution, mathematicians decided to invent a new number that would be the solution. "Let's call it **imaginary**," they said, having no trouble assuming all of their previous inventions were real. They decided to call this new number **i**.

---

b)   If we let $\mathbf{i} = \sqrt{-1}$, and we define $\mathbf{i}^2 = -1$, then what would be the value of :

$$\sqrt{-16} = \sqrt{16\mathbf{i}^2} = ?$$

CF-66.   Use the definition of **i** to show that:

a) $\sqrt{-4} = 2\mathbf{i}$ b) $(2\mathbf{i})(3\mathbf{i}) = -6$

c) $(2\mathbf{i})^2(-5\mathbf{i}) = 20\mathbf{i}$ d) $\sqrt{-25} = 5\mathbf{i}$

CF-67. In many cases we will have solutions that will be written in the form $a + bi$. (For example 1 - 2i). This number has a real component, the a part of the expression, and an imaginary component, the bi part. Numbers which can be expressed in the form of $a + bi$ are called **complex numbers**. Record this definition and the definition of **i** in your Tool Kit.

CF-68. In solving this system,
$$y = x^2$$
$$y = 2x - 5$$

we need to solve the equation $x^2 = 2x - 5$. Use the quadratic formula to find a solution which includes **i**.

CF-69. Simplify the following (write in $a \pm bi$ form):

a) $-18 - \sqrt{-25}$

b) $\dfrac{2 \pm \sqrt{-16}}{2}$

c) $5 + \sqrt{-6}$

CF-70. Explain why $i^3 = -i$. What does $i^4 = ?$

CF-71. Solve for x: $16^{x+2} = 8^x$

CF-72. Is $(x - 5)^2$ equivalent to $(5 - x)^2$? Explain briefly.

CF-73. Evaluate each of the following:

i) $(\sqrt{7})^2$ ii) $(\sqrt{18.3})^2$ iii) $(\sqrt{d})^2$

a)   Using what you observed above, evaluate: $(\sqrt{-1})^2$

b)   What does $i^2 = ?$

c)   Is this consistent with part (a)?

CF-74. Calculate each of the following:

a) $\sqrt{-49}$ b) $\sqrt{-2}$

c) $(4i)^2$ d) $(3i)^3$

CF-75. A function g(x) that "undoes" what another function f(x) does is called its inverse. In the last chapter, for example, we saw that $\log_2 x$ is the inverse of 2^x. We often use a special notation for this relationship and say that g(x) is actually $f^{-1}(x)$ (read " f inverse of x"). In other words, whatever f does to x, f^{-1} undoes.

a)   if $f(x) = 2x - 3$, then $f^{-1}(x) = ?$

b)   if $h(x) = (x - 3)^2 + 2$, then $h^{-1}(x) = ?$

CF-76. Consider a circle located at the origin with a radius of 5:

a)   If you make a quick sketch of this graph, how many points on the graph do you know are accurate? Explain.

b)   Find other points which will satisfy the equation $x^2 + y^2 = 25$. To get you started, consider (-4, 3). Look for a total of 12 points with integer coordinates.

CF-77. Another useful tool in finding an inverse function is a table. Consider the following table for $f(x) = 2\sqrt{x - 1} + 3$:

	1st	2nd	3rd	4th
what f does to x:	subtr. 1	$\sqrt{\ }$	mult. by 2	adds 3

Since the inverse must undo these operations, in the opposite order, the table for $f^{-1}(x)$ would look like:

	1st	2nd	3rd	4th
what f^{-1} does to x:	subtr. 3	div. by 2	$(\ )^2$	adds 1

a)   Copy and complete the following table for $g^{-1}(x)$ if $g(x) = \frac{1}{3}(x + 1)^2 - 2$:

	1st	2nd	3rd	4th
what g does to x:	adds 1	$(\ )^2$	div. by 3	subtr. 2
what g^{-1} does to x:				

b)   Write the equations for $f^{-1}(x)$ and $g^{-1}(x)$.

CF-78. Consider the following graph:

a)   What is the parent for this function?

b)   What are the coordinates of the locator point?

c)   Write an equation for this graph.

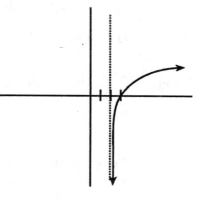

COMPLEX NUMBERS AS SOLUTIONS OF SYSTEMS OF EQUATIONS

CF-79.   When a graph crosses the x-axis, these x-values are sometimes referred to as the **roots** of the equation (in other words, when y = 0). We have seen that solutions to equations can be real or complex, so it must be possible for **roots** be either real or complex.

a)   Sketch the graph of $y = (x + 3)^2 - 4$. What are the roots?

b)   Sketch the graph of $y = (x + 3)^2$. What are the roots?

c)   Sketch the graph of $y = (x + 3)^2 + 4$. Find the roots by solving $(x + 3)^2 + 4 = 0$. Where does the graph cross the x-axis?

CF-80.   Show by purely algebraic methods that the graphs of y = 1/x and y = -x + 1 do not intersect. What are the complex solutions? Check the graphs on the graphing calculator.

CF-81.   Find the roots of each of the following quadratic functions by solving for **x** when y = 0. Do any of the graphs of these functions intersect the x-axis?

a.) $y = (x + 5)^2 + 9$.

b) $y = (x - 7)^2 + 4$.

c) $y = (x - 2)^2 + 5$.

CF-82.   What do you notice about the complex solutions of the equations in the previous problem? Describe any patterns you see. Discuss these with your group and write down everything you can think of.

CF-83.   Look for patterns as you find the products for parts (a)-(d). Then use your pattern to answer parts (e) and (f).

a) $(2 - i)(2 + i)$ b) $(3 - 5i)(3 + 5i)$

c) $(4 - i)(4 + i)$ d) $(7 - 2i)(7 + 2i)$

e)   Find a complex number by which you can multiply $(3 + 2i)$ to get a real number.

f)   Find a complex number by which you can multiply $(a + bi)$ to get a real number.

CF-84.   2 - 3i is 2 + 3i are called *complex conjugates*. What is the complex conjugate of:

a)   4 + i?   b)   2 + 7i?   c)   3 - 5i?

⌁CF-85.   Put examples of **complex conjugates** in your Tool Kit. Explain what happens when you multiply them and when you add them.

F-86.   Based of the following graphs, how many **real** roots does each polynomial function have?

i)   ii)

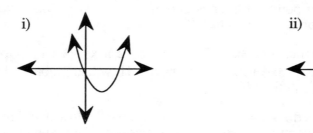   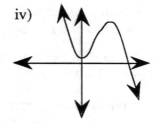

Something is added to each function and each graph is translated upward (resulting in the graphs below.) How many **real** roots does each of these polynomial functions have?

iii)   iv)

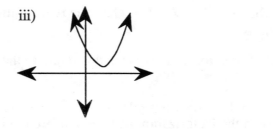

The polynomials in (iii) and (iv) do not have fewer roots. Polynomial (iii) still has **two** roots, but now the roots are complex. Polynomial (iv) has **three** roots, **two** are complex and only **one** is real.

CF-87.   Recall that a polynomial function with degree **n** crosses the x-axis at most **n** times. For instance, $y = (x + 1)^2$ intersects the x-axis once, while $y = x^2 + 1$ doesn't intersect at all. And $y = x^2 - 1$ intersects it twice.

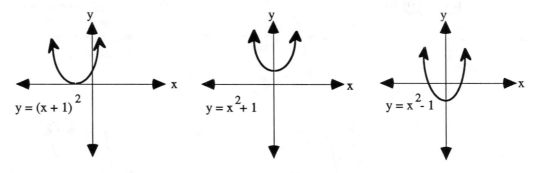

$$y = (x + 1)^2 \qquad y = x^2 + 1 \qquad y = x^2 - 1$$

a)   A third degree equation might intersect the x-axis 0, 1, 2, or 3 times. Make sketches of all these possibilities.

b)   Were all of the sketches possible? Why couldn't one of the sketches be done?

CF-88.   Now consider $x^3 - 3x^2 + 3x - 2 = 0$. Since the number of different real solutions is the same as the number of points of intersection of $y = x^3 - 3x^2 + 3x - 2$ with the x-axis, how many real solutions could this have?

   a)   Check to verify that $x^3 - 3x^2 + 3x - 2 = (x - 2)(x^2 - x + 1)$. How many real solutions does $x^3 - 3x^2 + 3x - 2 = 0$ have? How many non-real (complex)? What are all the solutions?

   b)   How many roots does $y = x^3 - 3x^2 + 3x - 2$ have? How many real and how many non-real (complex)? How many times does the graph intersect the x-axis? Check it out.

CF-89.   In our example with the parabolas, $x^2 + 2x + 1 = 0$ has one real solution, and $x^2 - 1 = 0$ has two real solutions.

   a)   What about $x^2 + 1 = 0$? Solve for x. How does this relate to the graph of $y = x^2 + 1$? Explain.

   b)   Consider the factors of the three polynomials: $(x + 1)^2$, $x^2 - 1$, and $x^2 + 1$. What is the relationship between the factorization and the number and kind of roots?

CF-90.   For each polynomial function f(x), the graph of f(x) is shown. Based on this information, tell how many linear and quadratic factors the factored form of its equation should have and how many real and complex (non-real) solutions f(x) = 0 might have. (assume a polynomial function of the lowest possible degree for each one)

EXAMPLE: 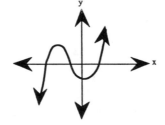   f(x) = 0 will have three real linear factors, therefore, three real solutions and no complex solutions.

a)   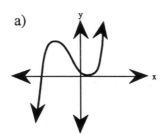   b)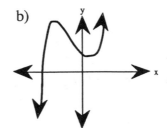

c)   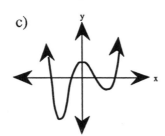   d)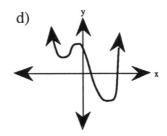

CF-91. Make a sketch of a graph f(x) so that f(x) = 0 could have the indicated number and type of solutions.

a)   5 real solutions.   b)   3 real and 2 complex.

c)   4 complex.   d)   4 complex and 2 real.

CF-92. For each part of the previous problem, what is the smallest degree each function could be?

CF-93. If $f(x) = x^2 + 7x - 9$, calculate:

a)   f(-3)   b)   f(**i**)   c)   f(-3 + **i**)

CF-94. Verify that 5 + 2**i** is a solution to $x^2 - 10x = -29$.

FINDING EQUATIONS FOR POLYNOMIAL FUNCTIONS

CF-95. Find reasonable equations for each of the following polynomial functions:

a)
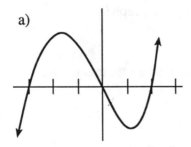

b)
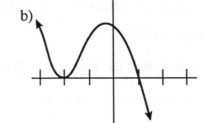

c)
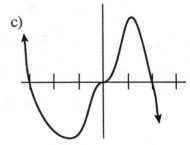

CF-96. What is the difference between the graphs of the functions $y = x^2(x - 3)(x + 1)$ and $y = 3x^2(x - 3)(x + 1)$?

CF-97. Find an equation for the graph to the right:

a)   If the equation were multiplied by 2, how would that alter the graph?

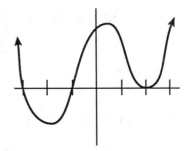

b)   The polynomial for this graph can be written as $P(x) = a(x + 3)(x + 1)(x - 2)^2$. Find the value of **a** if you know that the graph goes through the point (1, 16).

CF-98.   **THE COUNTY FAIR COASTER RIDE**

The Mathamericaland Carnival Company has decided to build a new roller coaster to use at this year's county fair. The new coaster will have the special feature that part of the ride will be under ground. They will use polynomial functions to construct various sections of the track. Part of the design is shown below:

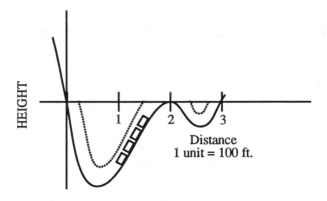

The numbers along the x-axis are in hundreds of feet. At 250 feet, the track is to be 20 feet below the surface. This will give the point (2.5, -0.2).

a)   What degree polynomial is demonstrated by the graph?

b)   What are the roots?

c)   Find the equation of the polynomial that will generate that curve.

d)   Find the deepest point of the tunnel.

CF-99.   Write a specific equation for each of the following graphs.

a)   b)

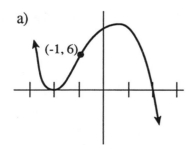   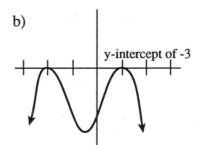

CF-100.   You are given the equation $ax^2 + bx + c = 0$.

a)   What are the roots of this equation?

b)   What part of the equation determines whether they are real or complex roots?

c)   When will the roots be real?

d)   When will the roots be complex numbers?

CF-101.   Consider the function $y = x^3 - 9x$.

   a)   What are the roots of the function? (Factoring will help!)

   b)   Sketch a graph of the function.

CF-102.   Which of the following equations have real roots and which have complex roots?

   a) $y = x^2 - 6$ b) $y = x^2 + 6$

   c) $y = x^2 - 2x + 10$ d) $y = x^2 - 2x - 10$

   e) $y = (x - 3)^2 - 4$ f) $y = (x - 3)^2 + 4$

CF-103. $y = \dfrac{1}{2}$ and $y = \dfrac{16}{x^2 - 4}$. Find the coordinates where the graphs of the functions intersect.

CF-104.   A circle is tangent to the lines $y = 6$, $y = -2$ and to the y-axis. What is the equation of the circle? Draw a picture!

CF-105.   Use algebra to solve the system of equations: $y = x^2 + 5$
          Confirm your solutions by sketching the graphs. $y = 2x$

CF-106.   You are given the equation $5x^2 + bx + 20 = 0$. For what values of b does this equation have real solutions?

CF-107.   The function h(x) is defined by the operations in the following table:

	1st	2nd	3rd
what h does to x:	adds 2	$(\ )^3$	subtr. 7
what h^{-1} does to x:			

   a)   Copy and complete the table for $h^{-1}(x)$.

   b)   Write equations for h(x) and $h^{-1}(x)$.

CF-108.   Find the center and radius of the following circles:

   a) $(y - 7)^2 = 25 - (x - 3)^2$ b) $x^2 + (y + 5)^2 = 16$

   c) $(x + 9)^2 + (y - 4)^2 = 50$ d) $y + (x - 3)^2 = 1$ (oops!)

CF-109.   In one of the games at the County Fair, people
pay to shoot a paint pistol at the target to the
right. The center has a radius of one inch. Each
concentric circle has a radius one inch larger
than the preceding circle. Assuming the paint
pellet hits the target randomly, what is the
probability it hits:

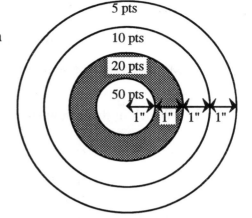

a)   the 50 point ring?

b)   the 20 point ring?

CF-110.   A device used for measuring distance on a playground or football field is a handle
attached to a rolling wheel which clicks with each full rotation. Brian needs to measure
the distance across his backyard, he sets the device at the start of a click and walks
across his backyard counting the clicks. He counts a total of 25 clicks. The radius of
the wheel is one foot. What is the distance across Brian's back yard?

THE GAME TANK LAB:
(CF-111 through CF-117)

CF-111.   The Mathamericaland Carnival Company wants to make a special game to premiere at
this year's county fair. The game will consist of a tank filled with ping pong balls.
Most are ordinary white ones, but there are a limited number of orange, red, blue, and
green prize ping pong balls. The blue ones have a prize value of $2.00, the reds,
$5.00, the orange ones, $20.00, and the green ping pong ball (there's only one!) is
worth $1000.00. There are 1000 blue ping pong balls, 500 reds, and 100 orange ones.
Fairgoers will pay $1.00 for the opportunity to crawl around in the tank for a set
amount of time, blindfolded, trying to find one of the colored ping pong balls.

The owner of the company thinks that she will make the most profit if the tank has
maximum volume. In order to cut down on the labor and materials costs of building
this new game, she hires you to find out what shape to make the tank.

The tank will be rectangular, open at the top, and will be made by cutting squares out of
each corner of an 8.5 meter by 11 meter sheet of transparent aluminum (this is the only
size available since the only other use for this material is whale aquariums).

Since it is difficult to cut and bend transparent aluminum (we can't get any, anyway!),
we will work with paper. You will need to write a report with your findings to the
Carnival Company. The report needs to include the following:

a)   The data and conjectures found in the experiment making the paper tanks.

b)   A well drawn diagram of the tank with the dimensions clearly labeled with
appropriate variables.

c)   A graph of the volume function that you found with notes on a reasonable domain
and range.

d)   An equation that matches your graph.

>>>PROBLEM CONTINUES ON NEXT PAGE>>>

e)   An estimate of the number of ping pong balls that will be needed and your adjusted recommendation based on this information.

f)   Your final conclusions and observations.

The following problems (CF-112 through CF-117) will help guide you through the various aspects of the project.

CF-112.   Use a sheet of paper the same shape as the material for the tank (at a slightly smaller scale!). Cut a square out of each corner. Measure the side of the square you cut out (you may choose the units you plan to use, but you will need to use these units consistently throughout the project). Cut a congruent square out of each corner. Your paper should look like this:

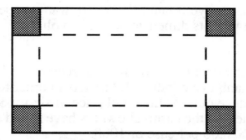

Now, fold it up into an open box (fold on the dotted lines). Then, using tape, tape the cut parts together so that the box holds its shape. Measure the dimensions of the tank . Record the dimensions directly on the model tank.

CF-113.   Make a table like the one below, calculate the volume of each tank and write it on the tank. Be sure to include proper units. You only need to collect eight or so different examples. Having each member of the group do two tanks would be an appropriate division of labor. Make sure that you consider "extreme" tanks, the ones with the largest possible cutout and the smallest possible cutout (do not actually do ones that would be physically impossible. You can't cut a square that is 1/100th of an inch on a side, but you can **imagine** cutting a square out of each corner zero inches on a side.)

Tank height	Tank width	Tank length	Tank volume

a)   Discuss with the other members of your group (experimenting if you wish) what happens if the cutouts are not square.

b)   Examine the data in the table from the different sizes of cutout squares and discuss it with your group. Make some conjectures about how to maximize the volume.

c)   Label the height as x. Using x for the height, find expressions for the length and the width.

d)   Find an equation for the volume of the tank.

e)   Sketch the graph of your function by using the roots and determining the orientation.

CF-114.   Graph the function on your calculator. Make sure you find appropriate values for the range on the graphing calculator.

   a)   Copy the graph onto graph paper with appropriate labels (what did 'x' represent?)

   b)   Find the maximum for the volume. And find the dimensions of the tank that will generate this volume.

CF-115.   Find maximums and minimums for the length, width, and height of the tank.

   a)   How are the dimensions of the tank related? In other words, what happens to the length as the height increases?

   b)   Make a drawing of your tank. (You may want to use isometric dot paper.) Label your drawing with its dimensions and its volume.

CF-116   The carnival company decides to follow your recommendation on the dimensions of the tank and to fill the tank to a depth of 1.2 meters in order to avoid losing the ping pong balls that might bounce out. At the retail price of six for $2.00 the ping pong balls would cost a fortune, but the carnival owners have found a real deal on army surplus ping pong balls at $20.00 per case of 1000.

   a)   A standard ping pong ball has a diameter of 3.7 cm. Estimate the number of ping pong balls they will need and the cost.

   b)   Oh, oh! The company realizes that they were overly ambitious (or greedy) in going for the maximum dimensions. They want to keep the tank the same shape, but smaller, and they ask you to calculate the reduced dimensions. They want to know whether taking the height of the tank down to 1.2 meters and keeping the level of the ping pong balls at a level 0.4 meters below the top of the tank will keep the cost of the ping pong balls under $8000. Work out the new dimensions for the tank, and present your conclusions as to whether they can fill the new tank with ping pong balls to a level of 0.8 meters for under $8000.

CF-117.   Based on your scaled down tank, what is the probability of winning a prize?

CF-118.   **PORTFOLIO: GROWTH OVER TIME PROBLEM**

          On a separate sheet of paper, explain **everything** that you now know about:

$$f(x) = 2^x - 3.$$

CF-119.   **GROWTH OVER TIME REFLECTION**

Look at your three responses to the growth-over-time problem. Write a self evaluation of your growth based those responses. Consider the following while writing your answer:

a)   What new concepts did you include the second time you did the problem? In what ways was your response better than your first attempt?

b)   How was your final version different from the first two? What new concepts did you include?

c)   Did you omit anything in the final version that you used in one of the earlier problems? Why did you omit that item?

d)   Draw some bars like the ones below and shade the amount that you knew when you did the problem at each stage.

First Attempt: ☐

Second Attempt: ☐

Final Attempt: ☐

e)   Is there anything you would add to your most recent version? What?

f)   Any additional comments about your mathematical growth this year.

CF-120.   **CHAPTER 7 SUMMARY ASSIGNMENT**

Identify with your group five main topics from this chapter. For each topic, rewrite or create a problem that represents the main ideas in that topic. Carefully solve and explain the solution to each of those five problems so that someone else, someone who didn't know what to do, could use these problems and solutions as example problems to help them through the chapter. This should be done on a separate sheet of paper to be included with your portfolio.

CF-121.   A polynomial function has the equation: $P(x) = x(x - 3)^2(2x + 1)$. What are the x-intercepts?

CF-122.   Sketch a graph of a fourth degree polynomial that has no real roots.

CF-123.   Sketch a graph of the following system: $x^2 + y^2 \leq 25$
                                                                                            $x - 2y > 5$

CF-124.   Consider the equation $x^2 = 2^x$:

a) How many solutions does the equation have?

b) What are the solutions?

CF-125.   A polynomial function has the equation $y = \mathbf{a}x(x - 3)(x + 1)^2$ and goes through the point (2, 12). Use the coordinates of the point to figure out what $\mathbf{a}$ must equal, and write the specific equation.

CF-126.   Graph the following system: $y = x^2 + 5$
                                                   $y = 2x$

      a)   Refer back to your algebraic solution to this problem from yesterday's assignment.

      b)   How does your graphical solution relate to your algebraic solution?

CF-127.   **COURSE SUMMARY ASSIGNMENT**

Identify with your group at least six, but no more than ten big ideas from units 1-7. After describing each big idea, rewrite or create one or two problems that illustrate the main point(s). Just as you did in the Chapter Summary, solve and explain the solution to each of those problems. This should also to be done separately from your homework and should be included with your portfolio.

The remainder of the problems are review.

CF-128.   Solve for x and y in each of the following systems:
These problems are best done with a graphing calculator. Remember that in order to use the calculator to graph the "xy =" parts of the first problem you have to write the equation in y-form.

      a) $2x + y = 12$ b) $y = -2x + 12$
                $xy = 16$ $xy = 20$

      c)   What is the difference between the graphs of the two systems (If you are not sure, you may want to graph them.)

CF-129.   Farmer Ted has been having more and more trouble with rabbits eating his crop. He grows alfalfa, and he has been losing more of his crop each month because of the growth of the rabbit population. Ted has been advised to have the rabbits destroyed, but he feels he can simply increase his production to compensate for the alfalfa destroyed by the rabbits. Currently Ted produces 600 tons of alfalfa each month and he plans to increase his production by 5 tons each month. Last month the rabbits destroyed 3 tons of the crop. The rabbit population around Ted's farm is increasing by 15% each month. If Ted allows the rabbits to continue destroying his crop, at what point will they eat everything he produces? Write a system of equations to represent his dilemma and estimate how long it will take.

CF-130.   A fifth degree polynomial has exactly two x-intercepts. What does this tell you about the roots of the function. Try sketching several different examples

CF-131. By looking at the graph of the function at the right:

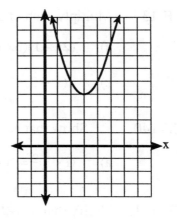

a)   what can you tell about the roots?

If the graph has a vertex at (3, 4) and it has a y-intercept of 13:

b)   what is the equation of the function?

c)   Find the roots of the function you found in part (b).

CF-132. If a system of equations has only complex number solutions, what do we know about the graphs of the equations?

CF-133. If $f(x) = \frac{1}{2}\sqrt{3(x-1)} + 5$, then $f^{-1}(x) = ?$

CF-134. Consider this geometric sequence: $i^0, i^1, i^2, i^3, i^4, i^5, ..., i^{15}$

a)   You know that $i^0 = 1$, and $i^1 = i$, and $i^2 = -1$. Calculate the result for each term up to i^{15}, and organize your answers so a pattern is obvious.

b)   Use the pattern you found in (a) to calculate $i^{396}, i^{397}, i^{398}$, and i^{399}.

CF-135. Use the pattern from the previous problem to help you to evaluate the following:

a) $i^{592} =$ b) $i^{797} =$ c) $i^{10,648,202} =$

CF-136. Describe how you would evaluate i^n where n could be any integer.

CF-137. The management of the Carnival Cinema was worried about breaking even on their movie ELVIS RETURNS FROM MARS. To break even they had to take in $5000 on the matinee. They were selling adult tickets for $8.50 and children's for $5.00. They knew they had sold a total of 685 tickets. How many of those would have to have been adult tickets for them to meet their goal.

CF-138. Sketch graphs of each of the following polynomial functions. Be sure to label the x and y intercepts:

a) $y = x(2x + 5)(2x - 7)$ b) $y = (15 - 2x)^2(x + 3)$

CF-139.   Multiply each of the following:

a) $(2x - 1)(3x + 1)$ b) $(x - 3)^2$

c) $2(x + 3)^2$ d) $(3 + 2x)(4 - x)$

CF-140.   Show that each of the following is true:

a) $(i - 3)^2 = 8 - 6i$ b) $(2i - 1)(3i + 1) = -7 - i$ c) $(3 - 2i)(2i + 3) = 13$

CF-141.   Verify that $x = -2 + 5i$ is a root of the quadratic function: $y = x^2 + 4x + 29$.

CF-142.   Carmel picks an integer from 1 to 12 as a value for **c** in the equation $y = x^2 + 4x + c$.
What is the probability that his equation will have complex roots?

CF-143.   A parabola has intercepts at $x = 2$, $x = -4$ and $y = 4$. What is the equation of the
parabola? There are several ways to solve this problem. Can you show more than one?

CF-144.   Solve each of the following:
(So that you can check, the answer to (b) is $4 \pm 3i$)

a) $x^2 - 4x + 3 = 0$ b) $x^2 - 8x + 25 = 0$

c) $x^2 - 4x + 20 = 0$ d) $x^2 + 100 = 0$

CF-145.   Write an explanation, including an example, of how to find out what f^{-1} does to x if
you know what **f** does to x.

CF-146.   Find the point where the graph of $y = 2^x + 5$ will intersect the line $y = 50$. Find your
answer to three decimal places.

CF-147.   Write equations for these circles:

a)   center (9, -3), radius 4   b)   center (-5, 0), radius $\sqrt{23}$

CF-148.   Write the equation for the circle with center (-4, 7) which is tangent to the y-axis.
Sketching a graph will help.

CF-149.   Today might be a good day to clean up and reorganize your Tool Kit. Suggestions for
"Building a Better Tool Kit" are included in the resource pages at the end of this chapter.

POLYNOMIAL FUNCTIONS LAB
(CF-1 & CF-3)

For each problem, complete the following:

a) Graph the function. Draw what you see in the window.
b) Use the **ZOOM** box feature to get a good view of how the graph curves. Sketch this view.
c) Use the **TRACE** feature to find the roots (x-intercepts) of the function. List the roots.
d) Shade on the number line where the value of the function (outputs) is positive.
e) Find the degree and describe the graph.

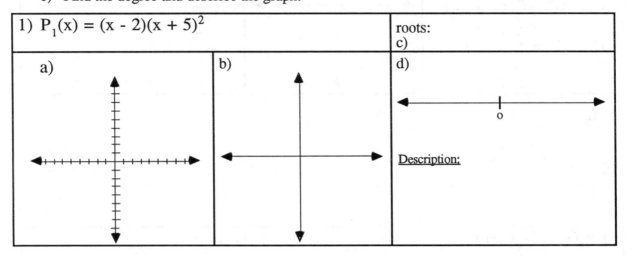

1) $P_1(x) = (x - 2)(x + 5)^2$

roots:
c)

a)

b)

d)

Description:

Notes:

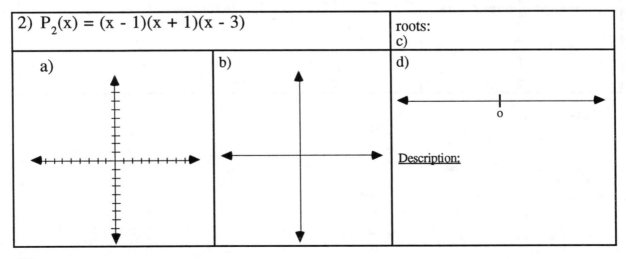

2) $P_2(x) = (x - 1)(x + 1)(x - 3)$

roots:
c)

a)

b)

d)

Description:

Notes:

3) $P_3(x) = 0.2x(x + 1)(x - 3)(x + 4)$

roots:
c)

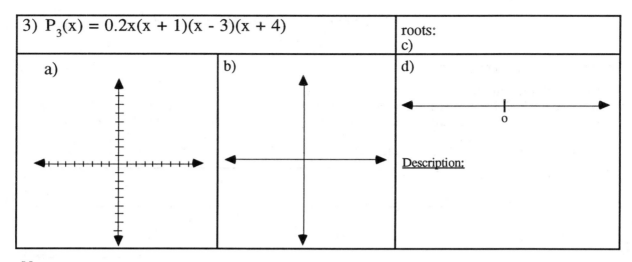

a)

b)

d)

Description:

Notes:

4) $P_4(x) = (x + 3)^2(x + 1)(x - 1)$

roots:
c)

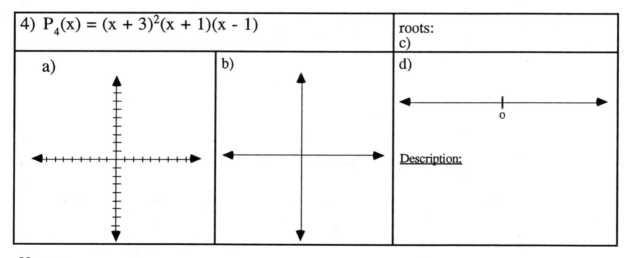

a)

b)

d)

Description:

Notes:

5) $P_5(x) = -0.1x(x + 4)^3$

roots:
c)

a)   b)   d)

Description:

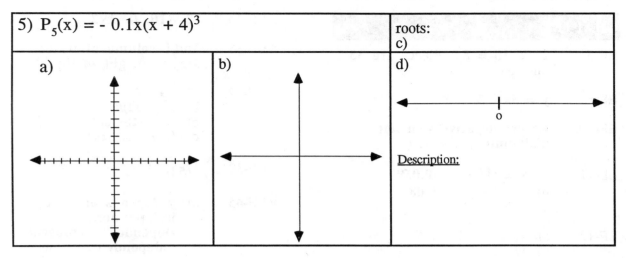

Notes:

6) $P_6(x) = x^4 - x^2$

roots:
c)

a)   b)   d)

Description:

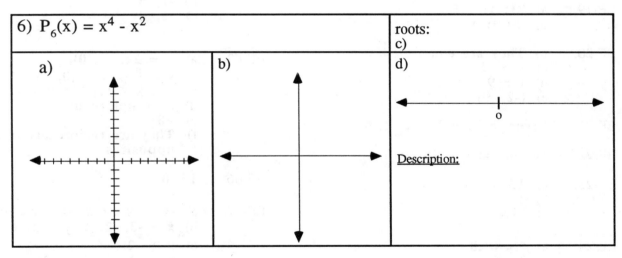

Notes:

CHAPTER 1

EF-5.   **2x + 18 = 84, Pieces are 33 and 51.**

EF-8.   **y = 5x - 2**

EF-10. **Error: negative sign not distributed, x = -37**

EF-11. a) $x = {}^{-1}/_{17} \approx -0.059$
        b) $x = {}^{66}/_{13} \approx 5.08$
        c) **x = -1, 3**

EF-12. a) **1, 2, 3, 4, 5, 6**
        b) $^1/_6$
        c) $^2/_3$

EF-18. a) **≈ 266.67**
        b) **37.5%**
        c) **27%**
        d) **135**

EF-19. a) **21; 15; 15**
        b) **-3; 3; 3**

EF-20. b) **They are equal.**
        c) **x = -2**
        d) **y = 9**
        e) **(-2, 9)**

EF-21. **Error in Line 2, x = -1, 5**

EF-22. **16 cm, 32 cm, 28 cm**

EF-23. a) **15.4**
        b) **1**
        c) **-13**

EF-24. a) $^4/_5 = .8$
        b) $^8/_5 = 1.6$
        c) $^{12}/_5 = 2.4$
        d) **y = $^{8x}/_{10}$ or y = ($^4/_5$)x or y = .8x**
        e) **y = 5.6**
        f) **Line passing through origin with, positive slope.**

EF-33. a) $x = {}^{5(y - 1)}/_3$
        b) $x = {}^{(-2y + 6)}/_3$
        c) $x = \pm\sqrt{y}$

EF-34. a) **6**
        b) **2**
        c) **0**
        d) **-4**
        e) **25**

EF-35. **-2, -10, 0, $\pm\sqrt{2}$, 5**

EF-36   **Ind.: volume of water, Dep.: height of liquid.**

EF-37. a) **19/24**
        b) **(a + 12)/(6a)**
        c) **(a + 2b)/(ab)**
        d) **((3d - 5c)/(cd))**

EF-39. $^1/_{52}$, $^{51}/_{52}$

EF-45. a) **y depends on x, x is independent.**
        b) **Dependent: temperature, independent: time.**

EF-47. a) **(10, 48)**
        b) **(-1, -5)**

EF-48. **Error: Line 3, x ≈ -4.39**

EF-49. a) **2, 5**
        b) **-7, 6**
        c) **-5, 0**
        d) **-1, 3/2**

EF-63. a) **y = 2x: (0, 0), y = $^{-1}/_2$x + 6: (0, 6), (12, 0)**
        c) **0 ≤ x ≤ 12, 0 ≤ y ≤ 4.8**
        d) **28.8 square units**
        e) **90°**
        f) **They are reciprocals and opposites**

EF-66. **1550**

EF-67. a) **-∞ < x < ∞; -∞ < y < ∞**
        b) **x = -3, -2, 2; y = 3, 1, -2, -3**
        c) **x ≥ -2; -∞ < y < ∞**
        d) **x = -3, -1, 0, 1, 3; y = -2, -1, 1, 2**
        e) **x = 2; -∞ < y < ∞**

PM-68. a) **x ≈ 781.36**
        b) **x = 6**
        c) **x = 0 or 1 or 2**
        d) **x = 1 or 1/5**

EF-69. a) **Find the radius or diameter**
        c) **2√(45π) ≈ 23.78 cm**

EF-74. **cube**

EF-75. a) **16**
        b) **9**
        c) **478.384384**

EF-76.   **0 and -2**

EF-77.   **Start by subtracting 5 from both sides of the equation so $x^2 - 10x + 16 = 0$, then $x = 8, 2$**

EF-78.   **$4w + w = 22$, $w = 4.4$, $l = 17.6$**

EF-79.
  b) **$y = \frac{2}{3} x - 2$**
  c) **Substitute $x = 0$ and solve for y. Substitute $y = 0$ and solve for x.**

EF-81.
  a) **$(5 \pm \sqrt{13})/2$; 3.45, 0.70**
  b) **$(-3 \pm \sqrt{21})/2$; -3.79, 0.79**
  c) **2, -3/2**
  d) **$(7 \pm \sqrt{193})/6$; 3.48, -1.15**

EF-82.   **1/26, 1/25**

EF-88.   b) **$\approx$ 400 miles.**

EF-89.
  a) **4**
  b) **103**

EF-90.   **$3x - 2y = 6$ or $y = (3/2)x - 3$**

EF-92.   **$x = 1, \; ^{-4}/_7$**

EF-93.   **18.11 cups**

EF-94.   **$y = 18.11 \, x$**
  a) **Independent : time (hours), dependent: cups of water.**
  b) **Domain: 0 to a large number, Range: 0 to some large number.**
  c) **Straight line, positive slope**

EF-95.   **$^n/_m$; $^m/_x$**

EF-100.   **478.384384**

EF-101.   **19937.44**

EF-102.   **x is independent in both, y and f(x) are dependent. They are essentially the same thing. $f(x) = y$.**

EF-103.
  a) **48.1131**
  b) **-20.078771**

EF-104.
  a) **(0, 3), $(^{-3}/_2, 0)$**
  b) **same as (a).**

EF-105.   **Only if $x = 0$.**

EF-106.   b) **b**

EF-107.
  a) **$D = -2 < x < 2$; $R = -3 < y < 2$**
  b) **$D = -2 < x < -1$; $R = -2 < y < 1$**
  c) **$D = -2 < x < 2$; $R = 1 < y < 2$**

EF-109.
  a) **$x = (y - b)/_m$**
  b) **$r = \sqrt{(A/_\pi)}$**
  c) **$W = V/_{LH}$**
  d) **$y = 1/_{(3 - 2x)}$**

EF-110.   **Graph 1: B; Graph 2: A; Graph 3: C**

EF-113.
  a) **-10**
  b) **-100**
  c) **-1000**
  d) **10**
  e) **100**
  f) **1000**
  g) **It's "blowing up." We cannot divide by zero.**

EF-116.
  a) **(3, 2)**
  b) **no solution**

EF-117.
  a) **1**
  b) **37**
  c) **6**

EF-118.   **43 hours**

EF-119.   **$\pm \sqrt{5}$**

EF-120.   a) **$z = {^{40}}/_3$**

EF-121.
  a) **-3**
  b) **-20**
  c) **12]**
  d) **1]**
  e) **0**
  g) **$x = 0, 1/2$**

EF-124.
  a) **$^2/_5$**
  b) **$^1/_5$**
  c) **Independent: colors, dependent: probability.**
  d) **Domain: blue, red, green. Range: $^2/_5, ^1/_5$**

EF-130.   **$m = {^{-3}}/_4$, (12, 0), (0, 9)**

EF-131.   **$5x - 3y = -15$**

EF-132.   **$\sqrt{34} \approx 5.83$**

EF-133. a) $8/27$
         b) $12/27$
         c) $6/27$
         d) $1/27$

EF-135. The ladder reaches $\approx$ 9.5 feet up the wall, so yes.]

EF-136. a) 1
         b) 12
         c) 13
         d) $\pm\sqrt{7} \approx \pm 2.65$
         e) $\pm\sqrt{26}/2 \approx \pm 2.55$]
         f) no solution]

EF-137. c) y-int (0,3) for both, x-int (-3/2,0) for (a), none for (b)
         d) (0,3) and (2,7)

EF-138. a) D: -2 < x < 1; R: -1 < y $\leq$ 3
         b) -2 $\leq$ x $\leq$ 2, -2 $\leq$ y $\leq$ -1

CHAPTER 2

BB-5.   b) Answers will vary.
        c) 1, 2, 4, 8, 16, 32, 64,...

BB-6.   a) $5^2 = 25$
        b) 3^{51}
        c) $(3 \cdot 4^4)/7$
        d) 6^{104}

BB-7.   a) y = -2x + 7
        b) $y = -3/2x + 6$

BB-8.   a) add 1, add 3, add 4
        b) 7, 15, 18
        c) 27, 75, 98
        d) t(n) = n + 2; s(n) = 3n; p(n) = 4n - 2

BB-9.   This is a scalene triangle.

BB-10. a) (x - 2)/(x + 2)
        b) (x - 3)/(2x + 1)

BB-11. (0, 0), (-6, 0)

BB-12. a) $1/4$
        b) $3/4$

BB-18. a) (-1, -2)
        b) (3, 1)

BB-19. soup $.79, tuna $1.39

BB-20. no, $m_{AB} = -1/5$; $m_{BC} = -1/3$; $m_{AC} = -1/4$

BB-22. a) y = -3x + 7
        b) $y = -x - 2/5$

BB-23. b) The soup cost $0.28 and the bread cost $1.08

BB-24. a) x: 0, 1, 2;   y: -2, 0, 1
        b) -1 $\leq$ x $\leq$ 1; -1 $\leq$ y $\leq$ 2
        c) -2 < x $\leq$ 2; -2 $\leq$ y < 2
        d) -2 $\leq$ x $\leq$ 1; -2 $\leq$ y $\leq$ 2

BB-25. a) 3
        b) $1/b$
        c) $1/b$

BB-26. 4

BB-27. a) $1/8$
        b) $3/8$ for exactly 2 tails; students might argue for the answer $1/2$

BB-38. m = 4, b = -5

BB-39. y= 4x - 5

BB-40. y = -3x + 9

BB-42. b) The lacrosse ball rebound is .625 - .681, and the handball rebound is .62 - .65.

BB-43. y = 3x + 10
        a) 24
        b) 64
        c) That a 3:1 ratio was used to adjust the points.

BB-44. $1/7$

BB-52. a) yes, 90th term
        b) no
        c) yes, 152nd term
        d) no
        e) no, -64 = n is not in the domain

BB-54. 10
        a) 248
        b) t(n) = 14n + 10

BB-56. a) geometric, mult by 1/2
        c) No, the sequence
           approaches zero. Half of
           a positive number is still
           positive.

BB-58. a) 5
        b) $4\pi/7$

BB-59. (0, -17), (-2 ± √21, 0) or
        (2.58, 0), (-6.58, 0)

BB-60. a) y = -1
        b) x = 0

BB-61. a) 1.03y
        b) .8z
        c) 1.002x

BB-62. a) 5, 7.5, 10, 12.5, 15, . . .
        b) 5, 5√2, 10, 10√2, 20 . . .

BB-63. a) 4n + 2; -12n + 24; 4n + 1
        b) yes, $2(3)^n$; $24(.5)^n$; $1(5)^n$
           or mult by 3, mult by 1/2,
           mult by 5
        c) Answers will vary.

BB-64. All domains: real numbers,
        ranges: y ≥ -1; y ≤ 1; y ≤ 0

BB-65. a) Pascal's triangle down to
           the fifth row.

BB-66. a) arith.; 3 t(n) = 3n + 1
        b) neither
        c) geom., r = 2
        d) arith.; t(n) = 7n + 5
        e) arith.; 1 t(n) = x + n
        f) geom., r = 4

BB-67. a) $3/4$
        b) $2/3$

BB-76. a) 5.3 feet to 5.8 feet
        b) 0.0175 ft to 0.0431 ft.
        c) $10(0.53)^n$ to $10(0.58)^n$ ft.

BB-77. $\left(\frac{14}{5}\right)$

BB-78. a) 1, 8, 15, 22, 29
        b) -5, -5, -5, -5, -5
        c) $1/16, 1/8, 1/4, 1/2, 1$
        d) 1, -2, 4, -8, 16
        e) whole numbers
        f) the result or sequence
           values

BB-79. a) -3, 1, 5, 9; Arithmetic;
           yes, the terms can be
           generated knowing the
           common difference and
           the initial value.
        b) Yes, it is term 81;
           let 4n - 3 = 321 and solve
           for n.

BB-81. a) .05x
        b) 1.05x
        c) Multiply by 1.07.
        d) 1.03 ; 1.0825 ; 1.0208

BB-82. a) 110
        b) 10100
        c) n(n+1)

BB-83. 17 marbles to start, add 13
        per year
        a) 13
        b) 94 years
        c) 17
        d) t(n) = 17 + 13n
        e) In 59 years.

BB-86. a) 1.05x
        b) 0.88x
        c) 1.0825x
        d) 0.775x

BB-87. a) 3252
        b) $1250(1.27)^n$
        c) guess & check to get 8
           weeks

BB-91. a) 285
        b) 56

BB-92. a) 0.20x
        b) .8x
        c) Multiply by .85

BB-93. a) .97
        b) .75
        c) .925

BB-95. a) 3
        b) 9

BB-96. a) 0, -50; 75, 125
        b) 25, 12.5; 50√2, 100√2
        c) t(n) = 100 - 50n,
           f(x) = 50 + 25x;
           $t(n) = 100(0.5)^n$,
           $f(x) = 50(\sqrt{2})^x$

BB-97.   Both have the same shape as $y = x^2$, one is shifted up 3 units, the other is shifted left 3 units.

BB-98.   a) 120
          b) 20100
          c) $n(n+1)/2$

BB-100.   In 14 days they will earn close to the same amount Plan A: $c(n) = 11.50n$; Plan B: $c(n) = .01(2)^n$

BB-101.   a) $455
          b) 63

BB-103.   a) i) they are equal ; ii) $z = 2y - x$
          b) i) z/y ; ii) y^2/x

BB-104. $1,396,569.60
          a) $575,918.47
          b) $s(n) = 673,500(1.20)^{n-1}$;
             $s(n) = 575,918(1.15)^{n-1}$
          c) $5,011,917; $3,883,058

CHAPTER 3

FX-9.   a) $x = 3$
          b) $x = 5$
          c) $x = 2$
          d) $x = 3$

FX-10.   a) zeroth
          b) just one, (2^0)

FX-11.   a) half of 2 is 1

FX-13. $(3 \pm \sqrt{65})/4$, or 2.77 and -1.27

FX-14..   a) 2^6
          b) 2^9
          c) 5^{2x}
          d) 2^{4x+4}
          e) $(2/3)^4$
          f) 3^8

FX-15.   b ; a
          a) $(4, -1)$
          b) $(-1, -2)$

FX-16.   a) x: $(6, 0)$ & $(-12, 0)$; y: $(0, -72)$
          b) x: $(4/5, 0)$; y: $(0,4)$

FX-17.   a) 1/2
          b) 1/2

FX-22.   a) 1.05
          b) 0.96
          c) $1.03^{12} = 1.426$
          d) $0.98^{12} = 0.785$

FX-23.   b) x^5, x^3
          c) x^{A-B}

FX-24.   a) $(0, 1)$
          b) $(0, 1)$
          c) $(0, 1)$
          d) $(0, 1)$

FX-25.   b) they do not
          c) undefined ; error

FX-26. $4^4 = (2^2)^4 = 2^8$ and $16^2 = (2^4)^2 = 2^8$

FX-27.   a) $x = 3$
          b) $x = 1$
          c) $x = 80$
          d) $x = 210$

FX-28.   a) 4
          b) -8
          c) 2

FX-29.   11/5 or 2.2

FX-30. $(3, 2)$

FX-39.   a) 1.06
          b) 24
          c) 368 years
          d) $V(t) = 24(1.06)^t$
          e) $24(1.06)^{368} = 4.93 \times 10^{10}$
          f) 3.4×10^{10}

FX-40.   a) 5^{-3}
          b) 3^{-2x}
          c) 2^{5x-5}

FX-41.   a) -3
          b) 3
          c) -2
          d) -3

FX-42.   a) 0.96; 126,000; 5 years
          b) 1.10; 0.35; 24 years in 1994
          c) 0.85; 11,000; 6 years

FX-43. $(2, -4)$

FX-44.   a) arithmetic
          b) $7 - (2/3)n$
          c) -3
          d) 42nd term

FX-45. no x-intercept; y: (0, -5)

FX-46. 3/9=1/3

FX-47. a) 3
        b) 2, -2
        c) -5

FX-58. a) 32
        b) 3125
        c) 2187

FX-59. a) x = 0.7
        b) x = 2/3
        c) x = 0.8

FX-61. c) yes
        d) yes
        e) yes
        f) yes

FX-62. all the same

FX-63. a) 1.0123 for 1 year
        b) 0.97 for one month
        c) $0.97^{12} = 0.694$ multiplier
           for one year

FX-64. a) x = 3
        b) x = 4
        c) x = -3/2

FX-66. z = xy/(y+ x)

FX-67. a) 5/10 = 1/2
        b) 3/10
        c) 2/10 = 1/5

FX-68. a) x = 0
        b) x = 0
        c) x = 0
        d) x = 1

FX-73. a) 275.5 million
        b) 336 million
        c) during the year 2055

FX-74. $2 = 1.04^x$; 18 years-solved
        by guess and check

FX-75. a) x = 7
        b) 103.82
        c) x = 9
        d) x = 1.5
        e) 1.75
        f) x= 3/2

FX-76. 8, 2^3, $(16^3)^{1/4}$, $(16^{1/4})^3$, and
        several √ forms

FX-78. a) 120
        b) 22,204

FX-79. a) x(x + 8)
        b) 6x(x + 8)
        c) 2(x + 8)(x - 1)
        d) 2x(x + 8)(x - 8)

FX-80. a) 5, 10, 17
        b) quadratic
        c) Discrete - It is a sequence

FX-81. $R_r = r_1 r_2 / (r_1 + r_2)$

FX-86. m = 1.118688942, or
        approximately 11.9% growth

FX-87. m = 0.8908987, so about
        11% decay

FX-89. a) x = 2
        b) x = 13/3
        c) x = 8/3

FX-90. a) 2(x+2)(x+2)
        b) 6(x+3)(x-4)

FX-91. a) 3
        b) 5
        c) 4
        d) 1/2
        e) 1/4
        f) 1/6
        g) 1/3
        h) 4
        i) a

FX-92. a) 3/2
        b) 3
        c) 6
        d) 2
        e) never ; 3

FX-94. a) x = 2
        b) x = 3
        c) x = -3

FX-95. a) y = (2x-7)/3
        b) y = (-1/2)x - 2

FX-96. a) x = -15
        b) x = ±5√2

FX-97. a) (-8, 2)
        b) (5/3, -1)

FX-98. (6, something)

FX-99. D = 3, F = 15, E = -1/3

FX-100. $(31)^2 = 961$

FX-108. $(24/7, 18/7)$

FX-109. a) 2, 6, 18, 54

FX-110. a) no solution
b) ≈ 3.17

FX-111. a) $x = 3, y = 2$
b) $x = 2, y = 1$

FX-112. a) $1/5^2 = 1/25$
b) $1/4^3 = 1/64$
c) $\sqrt{9} = 3$
d) $(^3\sqrt{64})^2 = 16$

FX-113. a) 5600
b) 5627.54
c) 5634.13

FX-114. $f(t) = 100(.87)^t$

FX-115. $(-60, -26)$

FX-117. x: $(-1, 0), (-13, 0), (0, 13)$,
$x = -7$

CHAPTER 4

PG-3. a) 4, 1, .25,
$t(n) = 256(.25)^n$
b) smaller
c) They get closer to zero.

PG-4. a) $y = (-2/3)x - 4$
b) $y = 2$
c) $x = 2$
d) $y = (2/3)x - 8/3$

PG-5. a) a cylinder
b) $45\pi \approx 141.37$

PG-6. a) $4\pi + (4/3)\pi \approx 16.755$ m^3
b) no - r, r^2, r^3 relationship.
Volume = $(80\pi)/3 \approx 83.776$ m^3
c) $V = (4/3)\pi r^3 + 4\pi r^2$

PG-7. $f(x) = x^2 + 1$

PG-8. a) $(2x - 3y)(2x + 3y)$
b) $2x^3(2 + x^2)(2 - x^2)$
c) $(x^2+9y^2)(x - 3y)(x + 3y)$
d) $2x^3(4 + x^4)$

PG-9. a)

n	t(n)	n	s(n)
0	0	0	1
1	3	1	2
2	6	2	4
3	9	3	8

b) Yes. The points lie on a line (since the sequence is arithmetic), so they have a constant slope, so the ratio of sides of triangles is the slope
c) no -- the points aren't collinear so the slope and therefore the ratios of the sides vary

PG-10. $(5, 14)$

PG-11. a) about \$182.50
b) $y = 150(1.04)^x$

PG-14. a) -3, 2
b) -18/5

PG-15. a) 0.71 hours or about 43 min
b) 0.91 hours or about 55 min

PG-16. a) Popcorn: \$3.75,
Soft drink: \$2.75

PG-17. a) $6x^4 + 8x^5y$
b) $x^{14}y^9$
c.) $(x + 3)/2$

PG-18. $x = (-by^3 + c + 7)/a$

PG-19. a) 8
b) $2a^2 + 16a + 32$
c) -7 or 1
d) -3

PG-21. a) $\sqrt{(61)}$ or 7.81
b) $\sqrt{(27)} = 3\sqrt{3}$ or 5.20

PG-27. a) $y = 0, 6$
b) $n = 0, -5$
c) $t = 0, 7$
d) $x = 0, -9$
e) They all have zero as a solution.

PG-28. a) $(7, -16)$
b) $(2, -16)$
c) $(5, -9)$
d) $(2, -1)$

PG-30. a) $y = (1/3)x - 4$
        b) $y = (6/5)x - 1/5$
        c) $y = (x + 1)^2 + 4$
        d) $y = x^2 + 4x$
        e) (a) x: (12, 0), y: (0, -4);
           (b) x:(1/6, 0), y: (0, -1/5);
           (c) x:(none), y:(0, 5);
           (d) x:(0, 0), (-4, 0), y:(0, 0)

PG-31. $(-23\pm\sqrt{561})/8$ and 0

PG-32. a) 7.656 gigatons
        b) $C(x) = 7(1.01)^{x+9}$ or
           $7.656(1.01)^x$

PG-33. 10.5 lbs. and 7.5 lbs.

PG-34. c) a circle

PG-35. a) 15 ft
        b) $528.76

PG-41. a) x: (1, 0), (-5/2, 0), y: (0, -5)
        b) x: (2, 0), y: none

PG-43. a) $g(1/2) = -4.75$
        b) $g(h + 1) = h^2 + 2h - 4$

PG-44. a) $x = \pm5$
        b) $x = \pm\sqrt{11}$

PG-45. $y = (-8/25)(x - 5)^2 + 8$
        standing at (0,0)

PG-46. a) A(n) is arithmetic, but
           a(n) is quadratic -- its rule
           is a parabola.
        b) A(n): line, a(n): parabola.

PG-48. a) $x = \pm\sqrt{y/2} + 17$
        b) $x = (y + 7)^3 - 5$

PG-49   a) 8
        b) ≈30.8

PG-50. b) loudness depends on
           distance

PG-51. a) i) 8; ii) 19; iii) 30;
           iv) not defined; v) x + y
        b) A ♣ B = A + B, unless
           A = B, in which case the
           rule is undefined.
        c) Factor the numerator and
           divide.

PG-58. $y \approx 2(x - 5)^2 + 2$ and
        $y \approx -1/2(x - 5)^2 + 2$

PG-59. a) $y = (x - 2)^2 + 3$
        b) $y = (x - 2)^3 + 3$
        c) $y = -2(x + 6)^2$

PG-61. x: (-10, 0), (8, 0),
        y: (0, -80) ; V: (-1, -81)

PG-62. a) $h(3) = 1/5$
        b) $h(-3) = -1$
        c) $h(a - 2) = 1/a$

PG-63. a) $x^2 - 1$
        b) $2x^3 + 4x^2 + 2x$
        c) $x^3 - 2x^2 - x + 2$

PG-64. x: (±1, 0), (2, 0),   y: (0, 2)

PG-65. x ≈ 21.14 so the other width
        is 528.39. There is another
        solution, but it is negative.

PG-66. a) $y = 100(1.08)^x$
        b) $342.59
        c) Domain: x ≥ 0,
           Range y ≥ 100

PG-71. for example: $y = (x - 4)^2$,
        $y = 5(x - 4)^2$, $y = -3(x - 4)^2$

PG-74. b) $y = 3x + 2$
        c) -1, 2, 5, 8, 11
        d) One is continuous and one
           is discrete. Same slope
           so "lines" are parallel but
           different intercepts.

PG-75. $x^2 + 2xy + y^2$

PG-76. a) 4.116×10^{12}
        b) $y = 1.665(10^{12})(1.0317)^t$

PG-78. a) x: (-3, 0); y: (0, 27)
        b) x: none; y: (0, 2)

PG-84. The second graph shifts the
        first 5 units left and 7 units
        up, and also stretches it by a
        factor of 4

PG-85. The second graph is a
        reflection of the first over the
        x-axis

PG-86. a) $y = 2x^2 - 4x + 6$
        b) none

PG-87. a) $y = x^3$
        b) $4\pi/3$
        c) cm^3
        e) Domain: $r \geq 0$; Range:
           $V(r) \geq 0$

PG-88. $y = .0l(x - 100)^2$ or
        $y = .01x^2$

PG-89. maximum profit is $25
        million when n = 5 million

PG-90. a) x: (-1/2, 0) (-1, 0); y: (0, 1)
        b) x = -3/4
        c) (-3/4, -1/8) or (-.75, -.125)

PG-91. move it up 0.125 units:
        $y = 2x^2 + 3x + 1.125$

PG-92. it's not a parabola anymore

PG-93. a) $y = x^2$
        b) no - not if a = 0

PG-100. The graph of x = 3 is the line
        of symmetry for the graph of
        $f(x) = | x - 3 |$

PG-102. a) 57, -43
        b) 43, -57
        c) -2, 22
        d) no solution

PG-103. none

PG-104. a) x = 4
        b) x = - 1
        c) x = -3
        d) x = 4

PG-105. a) x: (2, 0), (6, 0),
           y: (0, 2), locator or vertex
           (4,-2); Domain: all Reals;
           Range: y > -2, V: (4, -2);
        b) x: (-4, 0), (2, 0),
           y: (0, 2), locator or vertex
           (-1,3); Domain: all Reals;
           Range: y < 3, V: (-1,3)

PG-106. Move it up 6 units or redraw
        the axes 6 units lower

PG-107. ± 11, ± 9, ± 19

PG-108. a) y = x
        b) (1/2, 1/3)
        c) (1/2, 1/3)

PG-110. a) $8x^6y^3$
        b) $25x^2 + 5x + 0.25$
        c) $20s^2 - 245$
        d) -30375

PG-120. a) No
        c) $x = y^2$

PG-121. Brooke is right

PG-123. a) x: (-1, 0), y: (0, 2), V: (-1, 0)
        b) x: (0, 0), (2, 0), y: (0, 0), V: (1,1)

PG-125. $y = - (3/4)(x - 2)^2 + 3$

PG-126. a) y = 10.125
        b) x ≈ 6.18

PG-127. b) $15\sqrt{2} \approx 21.21$ cm

PG-134. a) yes
        b) (1, -1)
        c) D: all real #'s, R: y ≥ -1

PG-135. b) 28
        c) 2
        d) 1
        e) -1
        f) cube rt of -13
        g) no solution

PG-137. $y = (^{-5}/_9)(x - 3)^2 + 6$

PG-140. a) (0, -6)
        b) (-6, 0) and (1, 0)
        c) x: (0, 0), (-5, 0),
           y: (0, 0). The graph of p(x)
           is 6 units lower than q(x).

CHAPTER 5

LS-7.   a) y > 3x - 3
        b) y < 3
        c) $y \geq {}^{3x}/_2 - 3$
        d) $y \geq x^2 - 9$

LS-8.   a) Normal: C = 12h,
           High: C = 10h + 12
        b) Use High Use for more
           than 6 hours, Normal for
           less than 6 hours
        c) Same for 6 hours. It is
           the solution to the system.

LS-9.   a) ≈ 4 hours, Cost ≈ $73-74
        b) Cadillac = $92
        c) Cadillac = 9 hours and 10
           minutes

LS-10. c) **(-2, 3)**

LS-11. b) **$y = 0$**
c) **$x = 0$**

LS-12. a) **No solution**
c) **Parallel lines do not intersect.**
d) **Parallel planes**

LS-13. **Kamau will solve the system algebraically. Intersects at (2, 6)**

LS-14. a) **$x = 9/2$ $y = -3$**
b) **$x = -6$ $y = 4$**

LS-16.

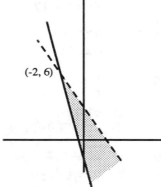

(-2, 6)

a) **B only**
b) **It shows where both inequalities are true**

LS-21. **$x = -2, y = 5, z = -\frac{1}{2}$**

LS-22. a) **$C = 800 + 60m$**
b) **$C = 1200 + 40m$**
c) **20 months**
d) **5 years**

LS-23. **26.5%**
a) **$x(1.05)^{10}$**
b) **$0.2x$**
c) **$0.2x(1.08)^{10}$**
d) **$0.2(1.08/1.05)^{10} = p/100$**

LS-25. c) **4N - 3, arithmetic**

LS-26. a) **$1/25$**
b) **x/y^2**
c) **$1/x^2y^2$**
d) **$(b^{10})/a$**

LS-27. a) **$1/27$**
b) **$2/9$**
c) **$4/9$**
d) **$8/27$**
e) **0**

LS-28. **The second one is shifted to the right 4 units and up 2 units**

LS-29. a) **$x \approx 36.78$**
b) **$x \approx 31.43$**

LS-30. **Rewrite as $(^{7}/_{2}) \cdot (4)^{-4}$ or $7/(2 \cdot 4^4)$**

LS-37. **No solution**

LS-38. a) **124**
b) **0**
c) **$n^3 - 1$**
d) **(2, 3, 3)**

LS-39. a) **input x, output x**

LS-40. **There were only 5 bacteria present.**

LS-41. a) **$5^{1/2}$**
b) **$9^{1/3}$**
c) **$17^{x/8}$**

LS-42. a) **square**
b) **(0,-3), (4,1), (-4,1), (0,5)**
c) **32 sq. units**

LS-43. **$x \approx 1.585$**

LS-44. a) **$1/11 \approx .0909$**
b) **1**

LS-48. **27 square units**

LS-49. **$x = -7, y = 11$**

LS-50. a) **1722**
b) **1368**
c) **$1500(1.047)^{n+3}$ or $1722(1.047)^n$**

LS-52. a) **$(b + a)x$**
b) **$(1 + a)x$**
c) **$a/(x + 1)$**
d) **$(x - b)/a$**

LS-55. **384 ft as the crow flies**

LS-57. **(-2, 3, -5)**

LS-58. **$a = 18.5, b = 5.5$**

LS-59. a) **$x^2 + y^2 = 36$**
b) **$(x - 2)^2 + (y + 3)^2 = 36$**

LS-60. $50(\pi/4) \div 400 \approx 9.8\%$

LS-61. $y = 0;\ x = 0$

LS-62. $y = -2x + 8$

LS-63.   a) 2^4
          b) 2^{-3}
          c) $2^{1/2}$
          d) $2^{2/3}$

LS-64.   a) $B = 0.07(0.3x)$ or $B = 0.021x$
          b) $S = 0.09(0.7)x$ or $S = 0.063x$
          c) $0.084x = 5000;\ \$59,524$

LS-65. $(x + 2)^2 + (y - 3)^2 = 4r^2$

LS-74.   b) Line parallel to y-axis at x = 4
          c) Plane parallel to yz-plane at x = 4
          d) point, line, plane.
          e) Point on x-axis.
          f) Line in xy-plane.

LS-75.   Red = 10 cm, Blue = 14 cm

LS-77.   No

LS-78.   (25, -3); new system (5,-3), (-5, -3)

LS-79.   Jonique is correct and Elissa is wrong

LS-80.   a) 0,-4
          b) 0, 2/5
          c) 0, 6

LS-81.   a) Approximately 1750.
          b) 4.14 lbs

LS-87.   a) $y = -2(x+4)^2 + 2$
          b) $y = 1/(x - 2)$
          c) $y = -x^3 + 3$

LS-88.   It is not the parent. The second equation does not have a vertical asymptote and it has a maximum value while $y = 1/x$ does not.

LS-89.   a) b/3
          b) b/(5a)
          c) b/(1 + a)}

LS-90.   The planes don't intersect in a single point; two or three parallel lines are formed or all the planes are parallel.

LS-91.   a) No. Input equals output only if $x \geq 0$

LS-92.   b) 6 square units

LS-93.   a) $2a^2 - 4$
          b) $18a^2 - 4$
          c) $2a^2 + 4ab + 2b^2 - 4$
          d) $2x^2 + 28x + 94$
          e) $50x^2 + 60x + 14$
          f) $10x^2 - 17$

LS-94.   a) $20.65
          b) $2.21

LS-96.   a) $1/5$
          b) $1/3$
          c) 100
          d) $(x + 60)/(x + 300) = 0.4$

LS-99.   a) $y = 2x^2 - 3x + 1$
          b) $y = (-1/2)x + 3$

LS-102. $y \leq -x + 4;\ y > 1/3 x$

LS-103.   (-1, -3, 7)

LS-104. $x = b/(1 - a)$

LS-106.   a) $1/4$
          b) $1 - x/12$

LS-107.   x = 3, y = 1, z = 3

LS-108.   a) -7
          b) 94
          c) 94
          d) 124

LS-109.   60°

LS-117.   b) 1, 3, 6, 10 ; No; No
          c) $0.5x^2 + 0.5x$

LS-119.   -1, 1/2, 2

LS-120.   a) $y = x^2 - 2x + 3$
          b) (1, 2)

CHAPTER 6

♀CC-5.
a) $y = (x-4)/2$
b) $y = (-3/2)x$
c) $y = 3x - 6$
d) $g(x) = \sqrt[3]{x - 1}$

CC-8.
a) 9
b) 4
c) $x \approx 1.89$

CC-9. They are equal, $x = y$.

CC-11.
a) $y = 2(x + 3)$
b) Yes, $y = x$.

CC-12. $x \approx \pm\, 9.5394$

CC-13. $x \approx 0.53$

CC-14. $(-3, 0, 5)$

CC-15. If she adds nothing else to the account and it just sits there making interest, she will have $440.13 on her eighteenth birthday.

CC-16. $x = 2.5$

CC-17. $(x + 3)^2 + (y - 5)^2 = 9$

CC-21.
a) $g(x) = \dfrac{x - 6}{8}$
b) $g(x) = \sqrt[5]{\dfrac{-x}{3}}$

CC-22. He will get the numbers 1, 2, 3, ... 50

CC-23.
a) f: reals, $y > 0$; f^{-1}: $x > 0$, reals
b) f: reals, reals; f^{-1}: reals, reals
c) f: $x \geq -3$, $y \geq 0$; f^{-1}: $x \geq 0$, $y \geq -3$,

CC-24. Trejo is correct.

CC-25.
a) $L(x) = x^2 - 1$; $R(x) = 3(x + 2)$
b) 30
c) Order does matter--show by plugging in numbers.

CC-26. 36

CC-27. 2

♀CC-28. Yes, order matters. Not necessarily equal.

♀CC-30.
a) $x^2 /(x-1)$
b) $(b + a)/(a - a^2 b)$

CC-32. ≈ 2.11 feet

CC-33. 1/5

♀LS-37. c) $y = a\sqrt{(x - h)} + k$

CC-38.
a) $e(x) = (x - 1)^2 - 5$
b) One machine undoes the other so $e(f(-4)) = -4$.
c) They would be reflections of each other through the line $y = x$.

CC-39. $f(x) = {(x - 7)}/{4}$

CC-40. No solution; planes don't intersect in a single point - parallel lines are formed

CC-42.
a) Temp. depends on time.
b) D: 0 to ∞, R: Room temp. to Starting temp., presumably boiling.
c) theoretically, yes; actually, no?

CC-43.
a) $x \approx \pm 5.196$
b) $x \approx 4.755$

CC-45.
a) 4, 20, 100
b) $n = 7$
c) No.

CC-52.
a) x: all numbers, $y > -3$
b) No.
c) $(0, -2)$, $(1.585, 0)$
d) $y + a = 2^x$, where $a \leq 0$

CC-53. $g(x) = (x^2 - 10)/5$; be sure the domain and range are properly restricted.

CC-54. Answer will vary.

CC-56. Yes, they are inverses since the x and y are just interchanged. No the second is not a function. Hence, the graphs are reflections of each other through the line $y = x$.

CC-57.
a) Quadratic.
b) $t(n) = n^2 - 3n + 4$

CC-58. n - 1; yes, n-1 is always one more\than n-2

CC-59. a) $x \approx 6.24$
        b) $x = 5$

CC-60. a) $(x - 1)^2 + y^2 = 9$
        b) $(x + 3)^2 + (y - 4)^2 = 4$

CC-65. a) $x = \log_5(y)$
        b) $x = 7^y$
        c) $x = \log_8(y)$
        d) $K = \log_A(C)$
        e) $C = A^K$
        f) $K = (1/2)^N$

CC-66. $.66

CC-67. $2x + 5y \leq 500, \ 6x + 3y \leq 450$
        no, because
        $6(60)+3(70)>450$

CC-68. a) Inflation.
        b) $t(n) = 1.5(1.048)^n$

CC-70. a) 16
        b) 12
        c) $12^4 = 20736$
        d) 54
        e) 3

CC-71. $h(k(x)) = k(h(x)) = x$

CC-72. a) ≈ 12353
        b) He will increase the size.

CC-73. $y = 2x \pm 2\sqrt{30}$

CC-74. 70

CC-75. $y = \log_3 x$.

CC-81. a) Base 12.
        b) Base 12 implies 12 fingers!

CC-82. $x \approx 17.673$

CC-83. $x = -4$

CC-84. a) $1/2 \leq x < \infty, \ 3 \leq y < \infty$
        b) $g(x) = \dfrac{(x - 3)^2 + 1}{2}$
        c) $3 < x < \infty, \ 1/2 \leq y < \infty$
        d) 6
        e) 6 - they are the same.

CC-85. a) True
        b) False - if $x = 2$ then $1 \neq 5/2 + 5/3$

CC-86. a) no solution because x cannot equal -4.
        b) $(-1 \pm \sqrt{5})/2$ or .61803 and -1.61803

CC-87. a) -0.889, -0.962, -0.988, -0.996
        c) The values get closer to -1.

CC-88. $m \approx 2.1867$

CC-89. a) $(b-a)/(b+a)$
        b) xy

CC-90. a) $1/4$
        b) $1/3$

CC-98. a) 25
        b) 2
        c) 343
        d) $\sqrt{3}$
        e) 3
        f) 4

CC-99. b) $a \log k = \log (k^a)$.

CC-100. $x \approx 17.6730$

CC-101. a) $(0,0)$
         b) $(0,0)$

CC-102. No; $\log_3 2 < 1$ and $\log_2 3 > 1$

CC-103. a) $k = y/(m^x)$
         b) m is the xth root of y/k

CC-104. a) The second is just the first shifted up ten.
         b) $y = k \cdot m^x + b$

CC-105. $y = a(x + b)^2 + c$ with $a < 0$ and $10 < c < 11$

CC-106. There are two ways you could restrict the domain, but usually people choose $x \geq -2$
         a) $y = \sqrt{((x+7)/3)} - 2$
         b) $x \geq -7; \ y \geq -2$

CC-112. a) 1.546
         b) 6.139
         c) 17.068

CC-114. take the 8th root

CC-115. a) $^1/_8$
b) $^1/_a$
c) m = 1.586...
d) 2.587 or -2.587
e) answers may vary, $x = b^{1/a}$

CC-116. x = 17

CC-117. a) x = -3, y = 5, z = 10
b) Infinitely many solutions
c) The planes intersect in a line.

CC-118. a) $x^{1/5}$
b) x^{-3}
c) $x^{2/3}$
d) $x^{-1/2}$
16.5 months; 99.2 months

CC-119. 0 < b < 1

CC-124. a) 5.717
b) 11.228

CC-129. a) (-0.75, 1.5)
b) (12,-7)

CC-130. a) Square it and subtract 5;
He dropped in a 76
b) $c(x) = x^2 - 5$.

CC-131. a) No.
b) Not necessarily.

CC-132. a) ≈ 0.0488 grams
b) ≈ 6640 years
c) Never!

CC-135. a) Decreasing by 30% means
we multiply by .7 each
time. Multiplier implies
exponential.
b) $y = 2350(0.7^x)$
c) $806.05
d) ≈3.83 years
e) $6156.06

CC-139. a) 1/2
b) any number except 0
c) 1.0×10^{23}

CC-140. a) 2.236
b) 4.230
c) 0.316
d) 2.021
e) 3.673
f) 3.659

CC-141. a) $g(x) = \sqrt[3]{(\frac{x}{3} - 6)}$
b) $f(x) = \sqrt[3]{(\frac{(x-6)}{3})}$
c) $f(x) = {}^{(x + 1)}/_{(x - 1)}$
d) $g(x) = {}^{(3x - 2)}/_x$

CC-142. a) ≈ $140,809.30
b) ≈ 24.2 years
c) ≈ $164,706.25

CHAPTER 7

CF-4. a) x = 2, 4
b) x = 3
c) x = -2, 0, 2

CF-8. a) 0 or 1
b) 0, 1 or 2
c) 0, 1 ,2, 3, or 4
d) 0, 1, 2, 3 or 4 - (1 and 3
require the parabola to be
tangent to the circle)

CF-11. c) not
f) not
g) not

CF-12. The second is up 5

CF-13. No

CF-19. at $x = (-3 \pm \sqrt{5})$

CF-20. a) 2
b) $(\sqrt{7}, 0,)$ and $(-\sqrt{7}, 0)$

CF-21. $x = -1 \pm \sqrt{6}$
a) 2
b) $(-1 + \sqrt{6}, 0)$ and $(-1 - \sqrt{6}, 0)$
c) at x ≈ 1.45 and x≈ -3.45

CF-23. at (74, 0), a double root, and
at (-29, 0)

CF-24. a) 2
        b) 5
        c) 3
        d) 6

CF-25. circle of radius 5, centered at origin, and its interior

CF-26. -1 or 5

CF-27. a) $y = (3^x) - 4$
        b) $y = 3^{(x - 7)}$

CF-29. The first statement is correct, but the second is false.

CF-37. a) Nowhere
        b) It has no real solution, therefore the graph cannot cross the x-axis.
        c) We get a square root of a negative.

CF-39. a) 13, 17, 21
        b) Arithmetic
        c) $4n + 1$

CF-40. a) $(x - 2)^2 + (y - 6)^2 = 4$
        b) $(x - 3)^2 + (y - 9)^2 = 9$

CF-42. area = 25 sq. units

CF-43. actually, it is not!

CF-44. a) $x = -26$
        b) $x = \dfrac{15}{4}$

CF-45. a) $4n - 23$
        b) 2506

CF-46. 20 days

CF-47. a) $2/20 = 1/10$
        b) $1/19$

CF-52. a) $x = 3$
        b) $y = 2$

CF-53. b) SC: $y = 20x + 180$, AH: $y = 80(1.15)^x$. They are equal in about 11.8 years

CF-54. a) 1/2
        b) 1/4
        c) 5/6
        d) 1/2
        e) 1/16
        f) $1 - \pi/4$

CF-55. Fred

CF-56. AC = 10 inches

CF-57. $-1 \pm \sqrt{2}$

CF-58. a) $y = 6x + k$ where $k \neq -7$
        b) $y = a(x - 4)^2 + 3$ where $a \neq 0$
        c) $y = a(x - 2)(x + 5)$ where $a \neq 0$

CF-59. circles concentric to (3, 5)

CF-60. circles with radius 3

CF-61. circles with radius 4, centers lying on a line 5 units above the x-axis

CF-68. $(1 + 2i, -3 + 4i), (1 - 2i, -3 - 4i)$

CF-69. a) $-18 - 5i$
        b) $1 \pm 2i$
        c) $5 + i\sqrt{6}$

CF-70. 1

CF-71. $x = -8$

CF-72. yes, both are $x^2 - 10x + 25$

CF-73. i) 7; ii) 18.3 iii) d
        a) -1
        b) -1

CF-74. a) $7i$
        b) $(\sqrt{2})i$
        c) -16
        d) $-27i$

CF-75. a) $(x + 3)/2$
        b) $\sqrt{(x - 2)} + 3$?

CF-76. a) four points: (5, 0), (0, 5), (-5, 0), (0, -5)
        b) (4, 3), (3, 4), (-3, 4), (-4, 3), (-4, -3), (-3, -4), (3, -4), (4, -3)

CF-77. b) $f^{-1}(x) = ((x - 3)/2)^2 + 1$; $g^{-1}(x) = \sqrt{(3(x + 2))} - 1$

CF-78.   a) log
         b) (2, 0), the x-intercept of
            the asymptote could be the
            locator point or the
            x-intercept, (3, 0) could be
            used as a locator point.
         c) $y = \log_2(x - 2)$ is one
            possibility

CF-84.   a) 4 - i
         b) 2 - 7i
         c) 3 + 5i

CF-86.   i) 2
         ii) 3
         iii) 0
         iv) 1

CF-88.   a) x = 2, $(1 \pm i\sqrt{3})/2$, one
            real, two non-real

CF-89.   a) $x = \pm i$, therefore it has
            no real roots and cannot
            cross the x-axis.

CF-90.   a) three real liner factors
            (one repeated), therefore
            2 real (1 single, 1 double)
            and 0 non-real roots
         b) one linear and one
            quadratic factor, therefore
            1 real and 2 complex
            (non-real) roots
         c) four linear factors therefore
            4 real, 0 non-real roots
         d) two linear and one quadratic
            factor , 2 real and 2 complex
            (non-real) roots

CF-92.   a) 5, b) 5, c) 4, d) 6

CF-93.   a) -21
         b) -10 + 7i
         c) -22 + i

CF-99.   a) $y = -2(x + 2)^2 (x - 2)$
         b) $y = (-3/4)(x + 2)^2(x - 1)^2$

CF-100. a) quadratic formula
         b) The root part
            (discriminant) - more
            specifically $b^2 - 4ac$
         c) If $b^2 - 4ac \geq 0$
         d) If $b^2 - 4ac < 0$ and $b \neq 0$

CF-101. a) (3, 0), (0, 0), (-3, 0)

CF-102. a) R
         b) C
         c) C
         d) R
         e) R
         f) C

CF-103. when $(x^2 - 4) = 32$, or
         x = ± 6, and y = 1/2

CF-104. $(x + 4)^2 + (y - 2)^2 = 16$ or
         $(x - 4)^2 + (y - 2)^2 = 16$

CF-105. (1 + 2i, 2 + 4i), (1 - 2i, 2 - 4i)

CF-106. b ≥ 20 or b ≤ -20

CF-107. b) $h(x) = (x + 2)^3 - 7$;
            $h^{-1}(x) = \sqrt[3]{x + 7} - 2$

CF-108. a) C:(3, 7), r: 5
         b) C:(0, -5), r: 4
         c) C:(-9, 4), r: $5\sqrt{2}$
         d) Parabola: V(3,1), down

CF-109. a) 1/16
         b) 3/16

CF-110. 50π ft ≈ 157 ft

CF-116   a) about 980,000 with the
            cost over $19,600
         b) The new tank will be
            1.2m x 5.85m x 3.98m.
            About 367,000 balls will
            fill it to a depth of 0.8 m
            at a cost of $7355

CF-117. about .004

CF-121. (0, 0), (3, 0) and (-0.5, 0)

CF-124. a) 3
         b) x = 2, x = 4 and x ≈ -0.767

CF-125. a = -2/3,
         $y = (-2/3)x(x - 3)(x + 1)^2$

CF-126. b) Since the graphs do not
            intersect, the system has
            no *real number* solution.

CF-128. a) (2, 8), (4, 4)
         b) (3 + i, 6 - 2i), (3 - i, 6 + 2i)

CF-129. $y = 600 + 5x$, $y = 3(1.15)^x$,
         In 40 months

CF-130. The function could have 3 real roots (one x-intercept must be a double root) and 2 complex roots. Or it could have one real triple root, one real double root, and zero complex roots. Or it could have one quadruple real root, another single real root, and no complex roots.

CF-131. a) The roots are complex numbers
         b) $y = (x - 3)^2 + 4$ or $x^2 - 6x + 13$
         c) $x = 3 \pm 2i$

CF-132. they do not intersect in the real plane

CF-133. $((2(x - 5))^2 / 3) + 1$?

CF-134. a) repeat 1, i, -1, -i, etc.
         b) 1, i, -1, -i

CF-135. a) 1
         b) i
         c) -1

CF-137. **450**

CF-138. a) x: -5/2, 0, 7/2;   y: 0
         b) x: -3, 15/2 (dbl root); y: 675

CF-139. a) $6x^2 - x - 1$
         b) $x^2 - 6x + 9$
         c) $2x^2 + 12x + 18$
         d) $12 + 5x - 2x^2$

CF-142. **8/12**

CF-143. $y = -0.5(x - 2)(x + 4)$ or
         $y = -0.5x^2 - x + 4$

CF-144. a) **1, 3**
         b) $4 \pm 3i$
         c) $2 \pm 4i$
         d) $\pm 10i$

CF-146. $\approx$ (5.492, 50)

CF-147. a) $(x - 9)^2 + (y + 3)^2 = 16$
         b) $(x + 5)^2 + y^2 = 23$)

CF-148. $(x + 4)^2 + (y - 7)^2 = 16$